LAND CHIEFS MINING

LAND
CHIEFS
MINING

SOUTH AFRICA'S NORTH WEST PROVINCE SINCE 1840

ANDREW MANSON AND BERNARD K MBENGA

WITS UNIVERSITY PRESS

Published in South Africa by:

Wits University Press
1 Jan Smuts Avenue
Johannesburg

www.witspress.co.za

First published 2014
ISBN 978-1-86814-771-7 (print)
ISBN 978-1-86814-772-4 (digital)

Project managed by Monica Seeber
Cover design and layout by Hothouse South Africa

FSC
www.fsc.org
MIX
Paper from
responsible sources
FSC® C105735 Printed and bound by Interpak

Table of contents

Acknowledgements

Many people, too numerous to mention, over the period of thirty years during which we have been studying and writing on the region's past, have afforded us invaluable insights and encouragement. We should like to single out Professor Sue Newton-King, formerly head of the Department of History at the University of Bophuthatswana (now North-West University, Mafikeng campus), who provided the start and enthusiasm required to embark on our respective paths as historians. We should also like to thank Jane Carruthers, Professor Emeritus in the Department of History, at the University of South Africa, who took the time to read and comment on a much earlier draft of this book. Her remarks helped tremendously in the final outcome of this project. Any shortcomings are entirely ours. We should also like to thank Eugene Breytenbach, Deputy-Director, Directorate of Information and Management, Office of the Premier, North West Province, and Liesl de Swardt, Department of Geography and Environmental Science, North-West University, Potchefstroom, for the production of maps. Our gratitude goes to Abe Madibo for photographs and to Paul Weinberg, Gille de Vlieg and Joe Alfers for the use of photographic material. And once again we are indebted to Elenore van der Riet of Hermanus for excellent translations of German scripts.

Terminology

We have chosen to employ the African prefixes when referring to the various communities about which we write. Thus we refer to the baHurutshe or baTswana in contrast to the academic norm which omits the prefix and retains only the stem: Hurutshe, Tswana. When used as an adjective the prefix 'ba' is omitted.

Glossary of Setswana and Afrikaans names

bakgosing	royal ward
commando	armed, mounted party
difaqane	period (c.1800 to 1830s) of political turbulence, migration and social transformation accompanied by frequent destruction of life and property in southern and central Africa
inboekeling (pl. inboekelinge)	indentured servant, likened to a slave
kgosana (pl. dikgosana)	clan head, sometimes referred to as a headman
kgosi (pl. dikgosi)	term for king or chief among all Tswana societies
kgotla	public meeting, central meeting place or court
khuduthamaga	advisory council to the *kgosi* comprising family members and *dikgosana*
laager	(Dutch/Afrikaans) defensive fortified position, usually circular, with use of wagons
makgowa	Europeans/whites
mephato	age regiments
morafe (pl. merafe)	chiefdom
oorlam(se)	Africans absorbed into Dutch/Afrikaner society and culture
pitso	public meeting
veldkornet	local district official (in the South African Republic) with administrative and, especially, military duties
voortrekkers	Dutch pioneers (later Afrikaners) who set out on the great trek from the Cape Colony from c.1834 to the 1840s and founded the South African Republic and the Orange Free State
volksraad	parliament of the South African Republic

MAP 1: North West Province: Location (2013)

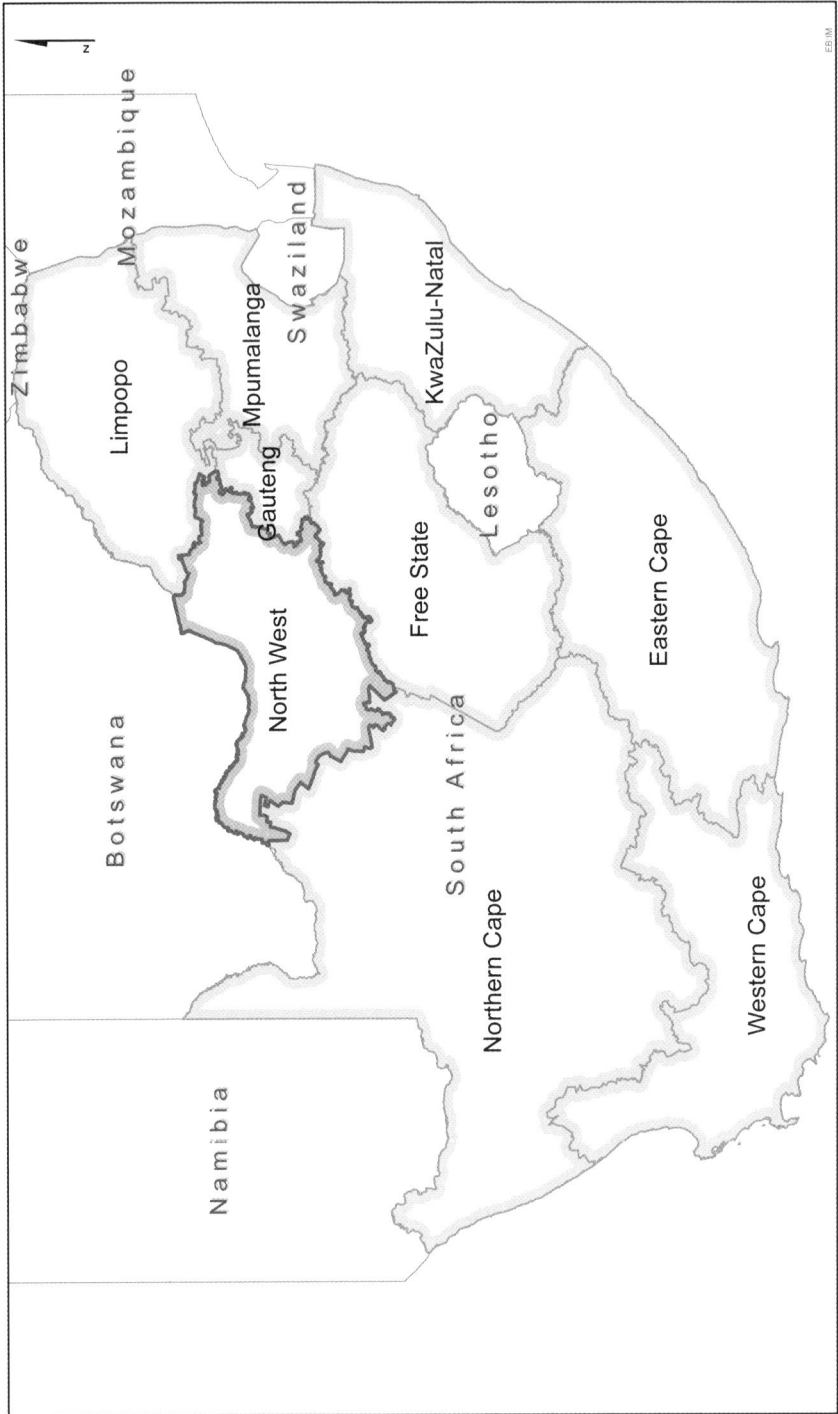

LAND, CHIEFS, MINING

MAP 2: North West Province: Bushveld Region – showing some towns, villages, roads, rivers and mountains (2013)

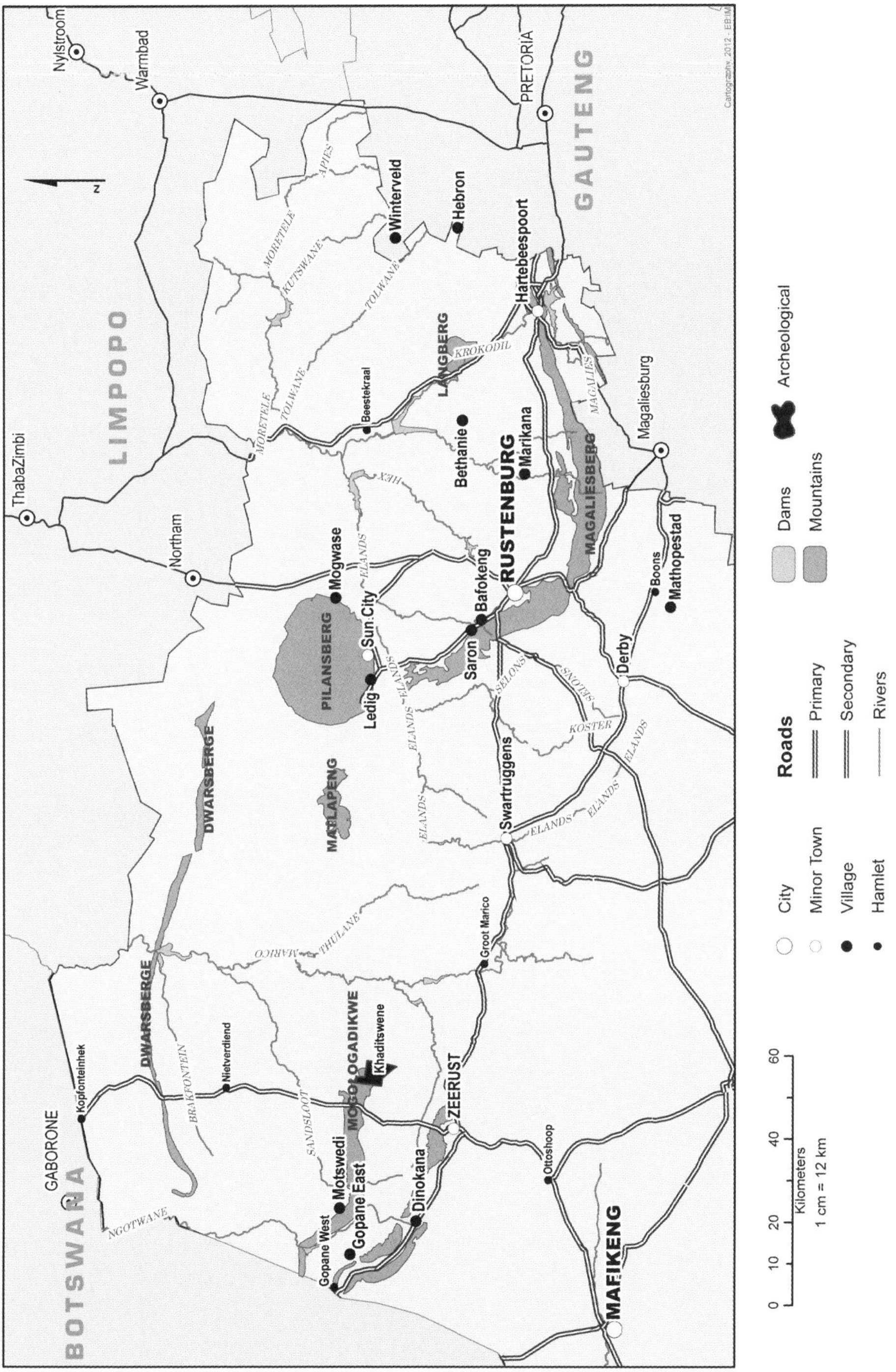

MAP 3: North West Province: Vryburg Region – showing some towns, villages, roads, rivers and dams (2013)

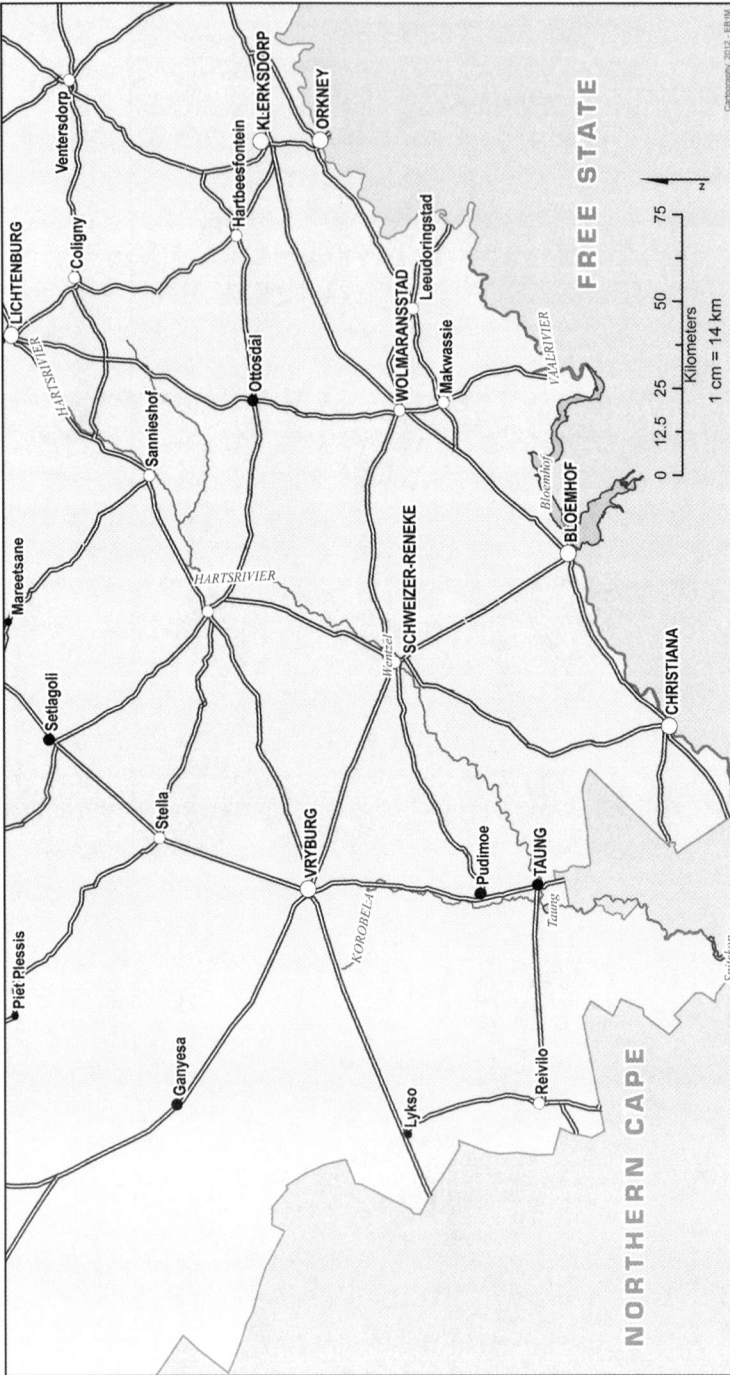

LAND, CHIEFS, MINING

MAP 4: North West Province: current ethnic areas (2013)

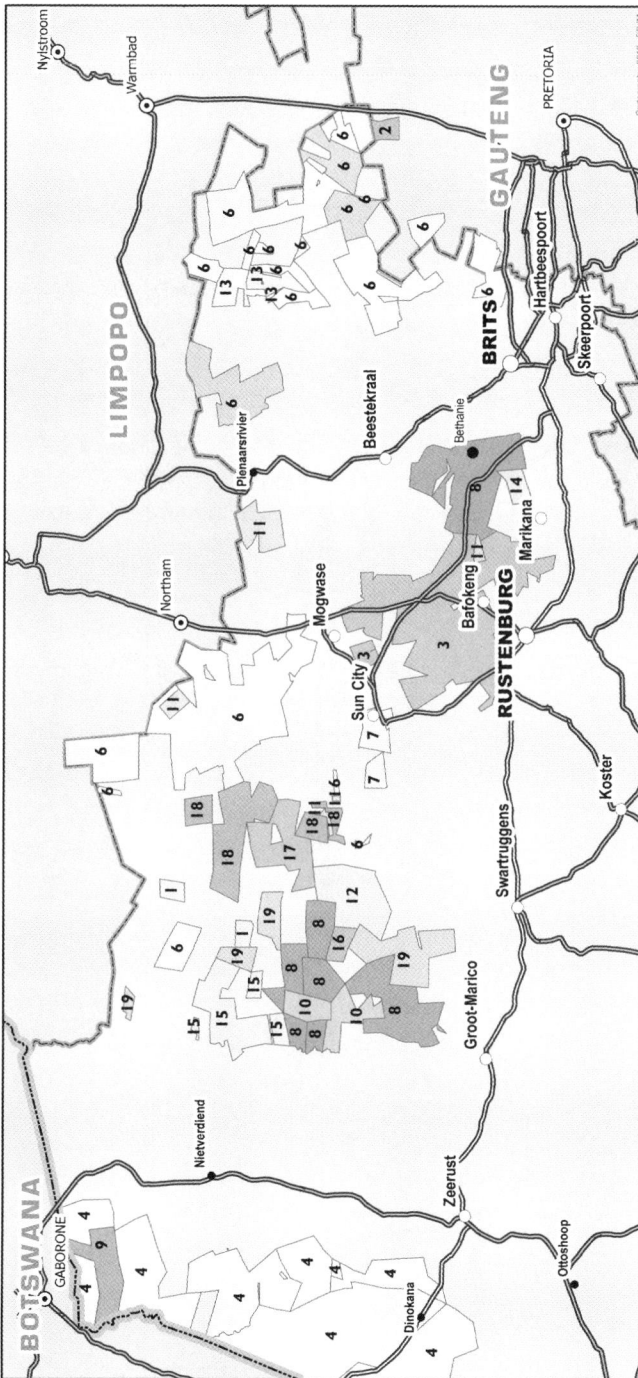

Table 1: Current Ethnic Areas

Area	Morafe	Area	Morafe	Area	Morafe	Area	Morafe
1	AMAHLUBI	6	BAKGATLA	11	BAPHALANE	16	BATAUNG
2	AMANDEBELE	7	BAHKUBUNG	12	BAPHIRING	17	BATLHAKU
3	BAFOKENG	8	BAKWENA	13	BAPHUTING	18	BATLHALERWA
4	BAHURUTSHE	9	BAMALETE	14	BAMALETE	19	BATLHOKWA
5	BAHWADUBA	10	BANURUTSHE	15	BAROKOLOGADI		

Roads
International Boundary

○ City
◎ Minor Town
● Village
● Hamlet

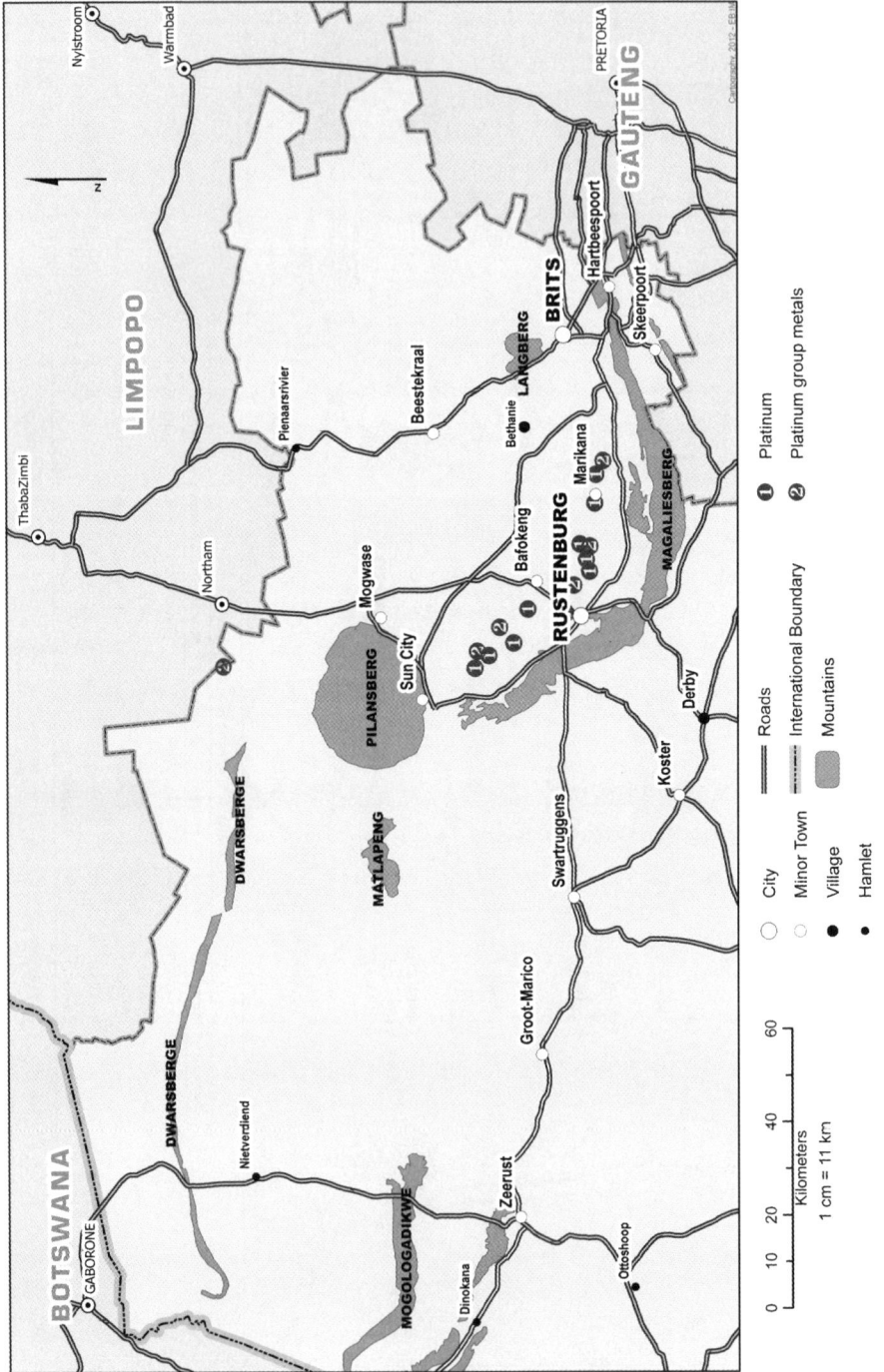

MAP 6: The Bechuanaland Reserves (2013)

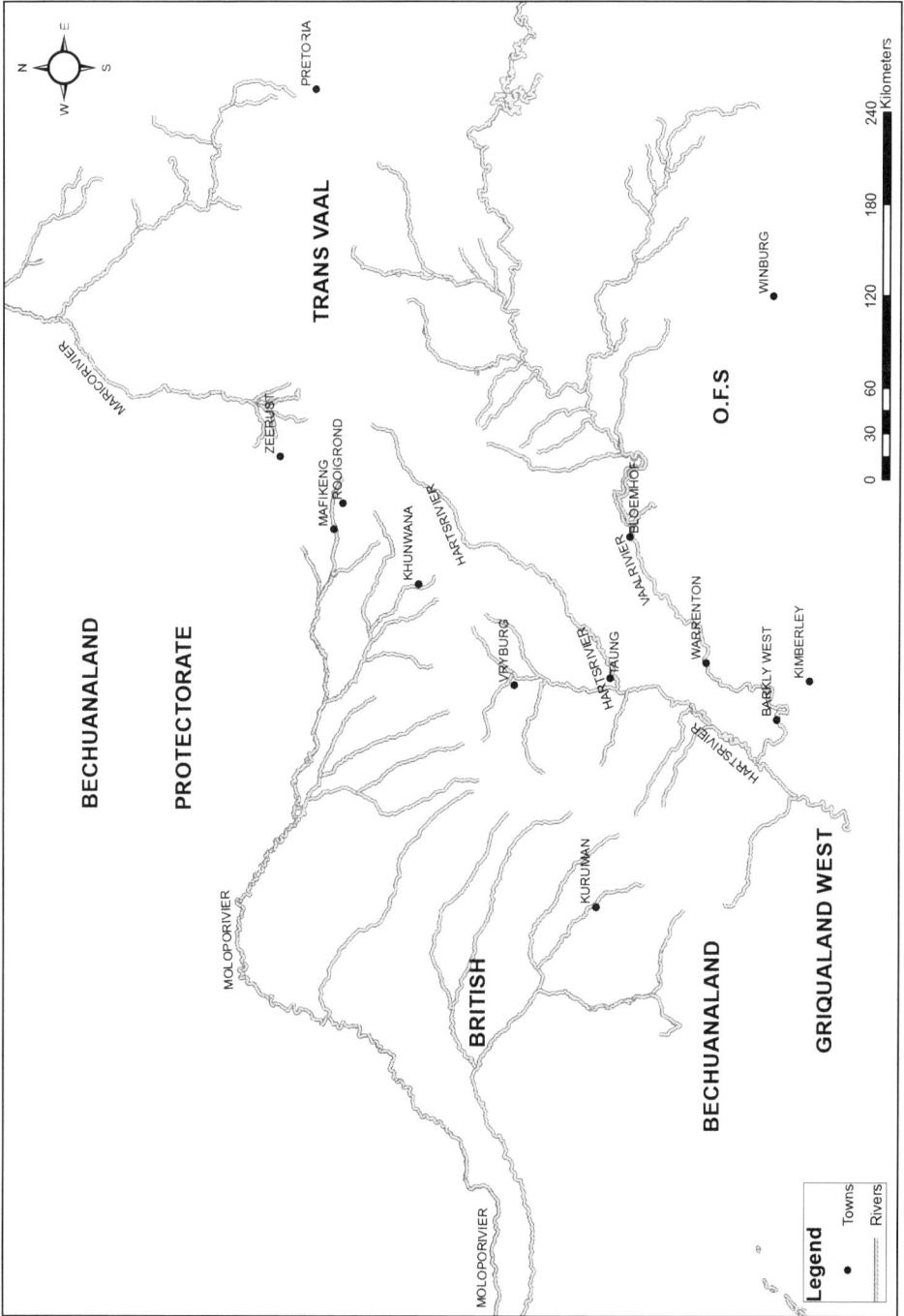

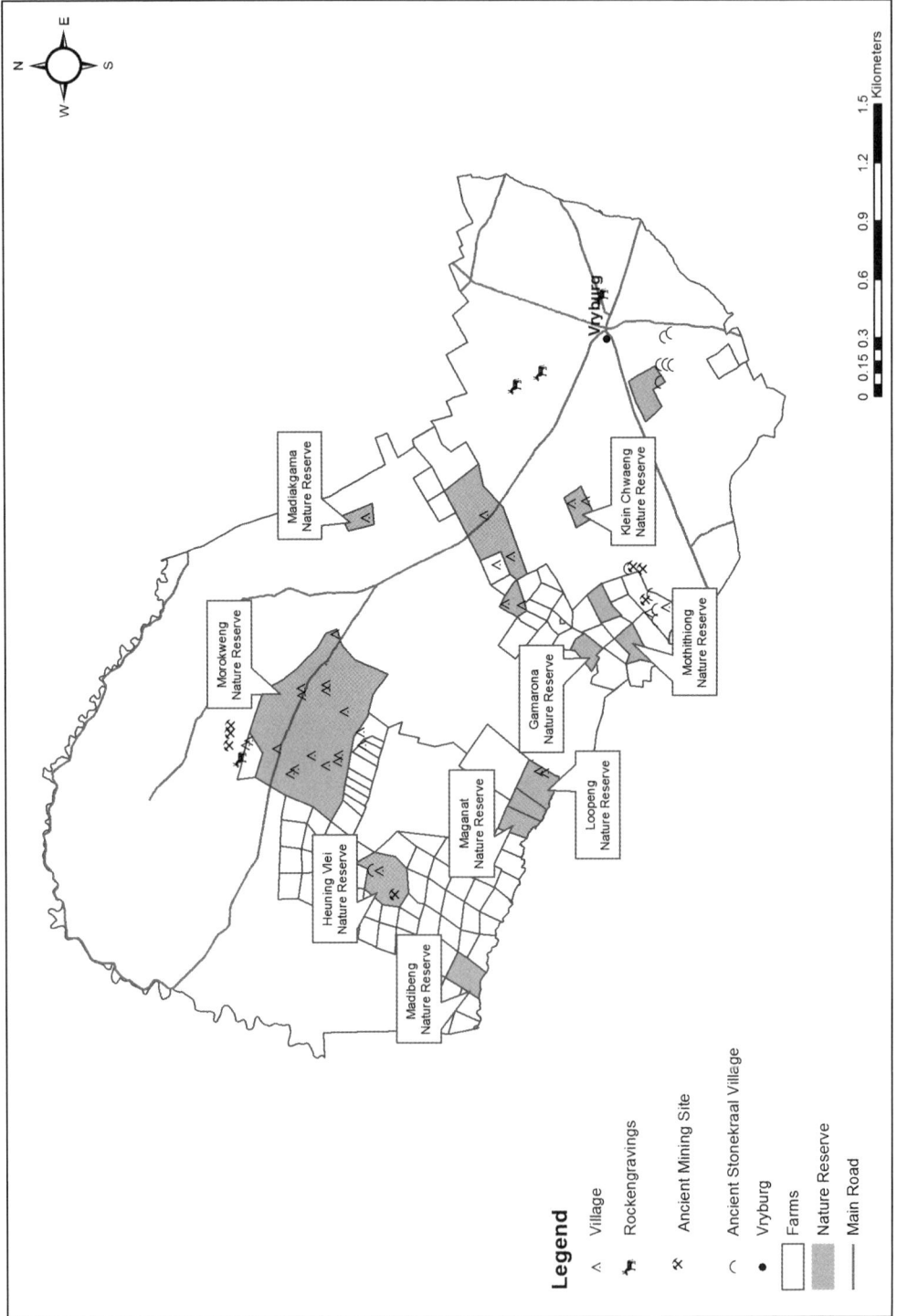

Legend

- ⋏ Village
- 🐂 Rockengravings
- ⚔ Ancient Mining Site
- ⦅ Ancient Stonekraal Village
- ● Vryburg
- ▢ Farms
- ▨ Nature Reserve
- — Main Road

Introduction

This book deals with aspects of the history of the black, predominantly Setswana-speaking population of today's North West Province of South Africa. It covers the period from approximately1840, with the beginning of settler and colonial domination, to the present. It is not a comprehensive account but, rather, a number of interrelated chapters on different topics which chart the various political and economic forces that have shaped the fortunes of communities and personalities in the province.

The North West Province is a recent geographical construct that arose out of the Constitution underpinning the new democratic dispensation in 1994. It comprises parts of the former western Transvaal, most of the former homeland of Bophuthatswana, and the northern reaches of the Cape Colony, later Cape Province (see Map 1). In one sense, the construct is not entirely artificial, for its inhabitants broadly comprise two culturally and politically homogeneous units – Setswana-speakers and Afrikaners – who have experienced close to 200 years of contact with one another. This is not to suggest that both societies were sealed off from outside influences. Both had extensive contact with their surrounding inhabitants and there was a constant infusion of other people into this region over a long period of time. Both societies interacted with British colonialism and bore the imprint of that association.

The history of the baTswana in South Africa has by no means been neglected. The early arrival of missionaries, traders and hunters from south of the Orange River, and the settlement of the Boers on the western highveld have ensured that many aspects of their societies were written down, providing a rich source of information for later scholars.

IMPORTANT PUBLISHED WORKS ON THE BATSWANA IN SOUTH AFRICA

Between the burgeoning of research and writing on African societies in South Africa beginning in the late 1960s, and its petering out some two decades later, the baTswana in the Republic – with the exception of Kevin Shillington's history of the colonisation of the Southern baTswana – rather missed the boat as far as published works are concerned. Shillington's work, however, principally covered the Southern Tswana living in the former colonies of Griqualand West and British Bechuanaland. As far as other Tswana chiefdoms are concerned, the baFokeng have been the focus of a recent study by the authors of this volume as well as Heinrich Baumann, and Fred Morton sheds light on events in the Pilanesberg district through several studies of the closely related baKgatla ba Kgafela in the Bechuanaland Protectorate.

Important aspects of twentieth-century Kgatla affairs are recounted in J Magala's history of the baKgatla ba Kgafela. Nancy Jacobs has written an environmental history of the black Tswana residents of the Kuruman district (although it is a little removed geographically from the North West Province). This declining attention to African societies in the pre-colonial and colonial eras was partly a reflection of increasing concern for other scholarly movements such as postmodernism, social and urban history, feminist and gender studies and, in South Africa especially, liberation histories which, on the whole, treated rural affairs in an understated way in which African reserves were viewed 'largely in terms of their functionality to the developing capitalist system'.[1] The other diversionary development was the rise of nationalism in Africa which focused on 'the larger narrative of national self-fulfilment'.[2] In this vision of Africa's past, colonialism was either regarded as dead and buried and best forgotten, or reformulated as neocolonialism and used as a justification for the failures of many modern African states. Recent times, however, have seen a shift in interest to the last two centuries in Africa, sparked by renewed interest in postcolonial and subaltern studies. Finally, recent interest in the impact of South Africa's 'bantustans' has led to a revival of interest in the lives of those trapped 'away in the locations'.[3]

Some of the more important books that have been published include Kevin Shillington's *The Colonisation of the Southern Tswana, 1870-1900* (Johannesburg: Ravan, 1985) which deals mainly with the baTlhaping and baTlharo *merafe* south of the Molopo River up to the turn of the nineteenth century. Part of this story has been reworked into a book on one of its leading figures, *Luka Jantjie: Resistance Hero of the South African Frontier* (London and Johannesburg: Aldridge and Wits University Press, 2011). Silas Modiri Molema wrote biographies of two prominent nineteenth century baRolong leaders, *Montshiwa 1815-1896, BaRolong Chief and Patriot* (Cape Town: Struik, 1966) and *Chief Moroka: His Life and Times* (Cape Town: Struik, 1950). The baFokeng received attention from the authors in *'People of the Dew': A History of the BaFokeng of Rustenburg-Phokeng Region of South Africa from Early Times to 2000* (Johannesburg: Jacana, 2010). Nancy Jacobs wrote a socioenvironmental history of the Kuruman district entitled *Environment, Power, and Justice: A South African History* (Cambridge: Cambridge University Press, 2003) that has implications for the wider thornveld districts of Mafikeng/Vryburg/Taung. For relations between the baKgatla in the Bechuanaland Protectorate and in the Pilanesberg see F Morton, *When Rustling Became an Art: Pilane's Kgatla and the Transvaal frontier, 1820-1902* (Cape Town: David Philip, 2009). An account of Kgatla affairs in the Transvaal is recounted in J Magala, *History of the Bakgatla baga Kgafela* (Crink City, 2009). Valuable, though in respects tainted, ethnographic information is available in P-L Breutz, *A History of the Batswana and the Origins of Bophuthatswana, A Survey of the Tribes of the Batswana, S Ndebele, Qwaqwa and Botswana* (Ramsgate: Breutz, 1989). M Legassick, *The Politics of a South African Frontier, The Griqua, The Sotho-Tswana and the Missionaries, 1790-1840* (Basler Afrika Bibliografen, 2010), (based on his doctorate of 1969), is an invaluable source for the late eighteenth/early nineteenth century. Of interest on other baTswana societies are C Murray, *Black Mountain, Land, Class, and Power in the Eastern Orange Free State, 1808s to 1980s* (Edinburgh: Edinburgh University Press, 1992; the cultural anthropological studies of the Comaroffs on the Tshidi-BaRolong – J Comaroff, *Body of Power, Spirit of Resistance: The Culture and History of a South African People* (Chicago: Chicago University Press) and J Comaroff and J Comaroff, *Of Revelation and Revolution: Christianity, Colonialism and Consciousness in South Africa* (Chicago and London: Chicago University Press, 1997) – which shed light on the processes of cultural diffusion and assimilation between the baRolong and the evangelising nonconformist mission movement in the nineteenth century and early twentieth century, closely bound up with the objectives of British colonialism.

NOTES

1   W. Beinart and C. Bundy, *Hidden Struggles in Rural South Africa: Politics and Popular Movements in the Transkei and the Eastern Cape*, 1890-1930 (Johannesburg: Ravan, 1987).
2   P. Limb, N. Etherington and P. Midgley (eds), *'Grappling with the Beast': Indigenous Southern African Response to Colonialism, 1840-1930* (Leiden and Boston: Brill, 2010). p.5.
3   *South African Historical Journal*, vol.64, no 1, (2012). See especially the Introduction by William Beinart, 'Beyond Homelands: Some Ideas about the History of Rural Areas in South Africa', pp. 5-21.

Generally speaking, however, the prominent Tswana personalities are less well-known and respected than leading figures among other African societies in South Africa such as the amaZulu, amaXhosa and baPedi, and some of their contemporaries in Botswana. Of course there are exceptions. Silas Modiri Molema wrote an excellent account of the life of Montshiwa of the Ratshidi baRolong, surely one of the most outstanding African figures of the nineteenth century.[1] Kevin Shillington published his important doctoral thesis on the colonisation of what he terms the Southern Tswana (mainly the baTlhaping, baTlharo and baRolong) in 1986, putting them on the map. His work culminates with the last act of local resistance, the Langeberg revolt in 1896 and its consequences: land confiscation, human displacement and immiseration. He afterwards wrote another book on the long-neglected personality Luka Jantjie, almost certainly the real hero of baTlhaping resistance to colonialism.[2]

In this volume we draw together previously unpublished material and existing literature, much of it resulting from our own research (but also that of others) to provide a fuller narrative of important aspects of the history of the Setswana-speakers and a few of its leading figures in the North West Province. A lot of this had been updated or re-interpreted. Only a portion of the scholarly work that has been conducted has been published, and much of it in rather esoteric publications not generally accessible or of interest to a general reading public. Our volume is an attempt to fill the gaps that exist in our understanding of the history of African people in the region, especially in the twentieth century, but this is not to imply that the collection represents the final or complete word on the region's past. It is intended more to open up a number of perspectives, and a subset of shared experiences, on the history of this particular region of South Africa. We hope to expand our understanding of Setswana-speaking communities in South Africa; to inform and enthuse students of South African history; and to attract a wider readership among those whose pasts we recount.

Setting the scene: The land and its inhabitants.

Geographically, one can broadly divide the North West Province into two subregions. The first is the western bushveld which stretches from the western Magaliesberg to the Marico district. The bushveld is not a neatly bounded region (literally, it describes a form of terrain and vegetation characterised by quite dense woodland and tall grasses) but lies between the southern Kalahari Desert and the land slightly west of the Magaliesberg range. More specifically, within this is the area bounded by the Madikwe/Ngotwane rivers in the west, the Limpopo River in the north and the Odi (Crocodile) river system, comprising the Elands (Kgetleng), Apies (Tshwane) and Pienaars (Moretele) rivers, in the east. The water flow from the Odi river system drains into the Limpopo. Not all these rivers are perennial. Today this region forms the northern part of the North West Province where it borders on the Limpopo Province and the Republic of Botswana. Three mountain ranges punctuate the generally undulating nature of the region: the Dwarsberg (Motlhwane) in the north-west, the crumpled ridges of the Swartruggens in the centre, and the Pilanesberg crater in the south-east. Human habitation, in the past and even at present, becomes more scattered and sparse the further north one proceeds, until the higher plateau of the Waterberg in Limpopo affords a milder and more pleasant environment (see Map 2).

The western bushveld was largely impenetrable for Africans and Europeans alike in the nineteenth century. One intrepid traveller, Adolphe Delegorgue, attempted to cut through the bushveld in the early 1840s. Within a week he had begun to turn back, forced by the 'haak-doorn (*acacia*) whose fang-like thorns tore pitilessly at the flesh … like fishhooks', the 'death of the last of my oxen … from the sickness which people attribute to flies', and the insects (specifically ticks and mosquitoes) which tormented him 'as no others had in Africa'.[3] The unfortunate man was furthermore plagued by a tapeworm that forced him to consume a vast amount of meat each day.

Wildlife was still plentiful in the early nineteenth century, supporting early occupants of the region who were adept hunters, and attracting the early attention of gun-using hunters on the western highveld. Most of the animals of the African savannah were to be found here and lions were a common scourge. John Campbell, a missionary of the London Missionary Society (LMS) in 1821, noticed that the African population was forced to build special elevated sleeping structures for their children to protect them from lions,[4] and the explorer David Livingstone nearly met his end in a lion attack in 1846 in the Gopane area, just north of modern Zeerust. Well-known nineteenth-century big-game hunters such as Gordon Cumming, William Cornwallis Harris and Frederick Courtney Selous found the area to their liking. Visiting the Pilanesberg in 1836, Harris remarked on the sight of 'three hundred gigantic elephants, browsing in majestic tranquility amidst the

FIGURE 1: The road to Dinokana c. 1987 – typical bushveld terrain

Source: Joe Alfers

wild magnificence of an African landscape, and a wide stretching plain darkened, as far as the eye could see, with a moving phalanx of gnoos and quaggas whose numbers simply baffle computation'.[5]

The second subregion comprises the area south of the Molopo River (bordering on Botswana) to the southern reaches of the Harts River before its confluence with the Vaal, and east of the Ghaap plateau: that is, the land between modern Mafikeng down to Vryburg and Taung. This is a drier and ecologically more limited part of the province. Though criss-crossed with a number of westward-flowing river systems, the river beds remain dry for most of the year. With an average rainfall of between 38 and 56 cm agriculture provided only a precarious source of livelihood, particularly in times of drought. In the early to mid-nineteenth century the vegetation comprised mainly intermittent but palatable grasses, bushveld scrub, trees and succulents, usually classified as 'Kalahari thornveld'. This grassland could sustain cattle and other animals, especially in the Molopo basin where more rain fell and the grasses were sweeter. Fortunately, for the inhabitants and animals alike, limestone outcrops give rise to many springs or fountains (also called 'eyes') that provide constant water, and enable boreholes to be sunk, especially close to the river beds. Since the mid-nineteenth century, overcrowding and ecological degradation has caused the region's flora to deteriorate. In his pioneering book *The Road to the North*, the historian AJI Agar-Hamilton describes this part of the country as 'bleak, arid and treeless', and exposed to 'a desiccating dry wind'.[6] However, early nineteenth-century observations by travellers, missionaries and naturalists, most of whom pioneered the hunters' and missionaries' 'Road to the North', suggest that it was more arable and provided better conditions for human habitation and the fauna of the region than it does today (see Map 3).

Until well into the nineteenth century, this territory accords with what historians have described as an 'open' frontier, one where two distinct societies, one indigenous (in this case the Setswana-speakers) and one intrusive, comprising white adventurers and other immigrants (principally the Griqua and Kora), encountered one another and struggled to establish full hegemony over the region. This north-west bushveld frontier opened up as the Transorangia frontier further south closed in the 1840s. The Transorangia region had, from the late eighteenth century, been a 'mélange of people': Griqua, Kora, Khoisan, Sotho-Tswana and a handful of white farmers, for the most part seeking stability in a volatile region.[7] Some of the inhabitants of the north-west frontier, through trade activities and by displacement from the bushveld during the migratory years after the so-called *difaqane* in the 1820s, had in fact already experienced life in Transorangia.

The north-west frontier closed in two stages, first with the declaration of the Bechuanaland Protectorate in the 1880s (which led to British control and safe passage through to the Ndebele state in Matebeleland), and second with the Boer defeat of Mabhogo during 1894 (which provided the Boers with easier occupation of the Waterberg district) and the rinderpest epidemic (1896) which killed off much of the wildlife hosting the tsetse, thereby facilitating trekker penetration of the Limpopo valley.[8] As Neil Parsons observes, 'contemporary maps customarily marked the Limpopo as the boundary (of the South African Republic, SAR) and South African historians have accepted this fiction as if before the 1900s the SAR had indeed "filled out" as far north-west as the Limpopo'.[9] The failure of the South African Republic accurately to define its borders or even to publish official maps led to numerous disputes between Africans and the trekkers or, as the *Transvaal Argus* of 1876 put it, 'perpetual haggling and bandying of words with a dozen (native) chiefs'.

The sense of region is not carried forward into this narrative in any intentional thematic way. However, this is as much a 'regional' history as anything else, one in which the people happen to share a common past and are mainly Setswana-speakers. The territory offers a number of unique features and evokes quite specific images in the popular imagination. Most obvious and most current is its association with mining. It is home to what is called the Bushveld Igneous Complex (BIC) which, along with Zimbabwe's Great Dyke Complex, is in turn home to the largest concentration of the ore-bearing lodes that contain the Platinum Group Metals (PGMs). The metals found in the BIC are distinguished by the fact that they contain more platinum than other areas in the world where the PGMs are mined. But in addition to this the BIC contains the world's largest reserves of chrome and vanadium, both part of the PGM. At the heart of the BIC is the Merensky Reef whose core runs through the western bushveld. Global demand and improved technological ability led to a steady increase in platinum mining from the early to mid-1980s to the point where 44 million ounces of the metal were refined in 2011. The mining of PGMs constitutes what can be termed the third mineral revolution in South Africa, after diamonds and gold. The earlier revolutions led to the expansion of colonialism and the rise of the capitalist state, at considerable cost to the indigenous population; the platinum revolution coincided with the advent of democracy, with different interest groups able to exert pressure on the state. This has introduced a new set of complexities and contradictions that are still playing themselves out in the second decade of the new millennium.

The consequences of this unbridled explosion of mining in the BIC led to the migration of large numbers of workers into the region, most of them ethnic 'strangers' from other parts of South Africa. Low wages and poor living conditions on the mines led to bouts of labour unrest that culminated in the strike at Lonmin's Marikana plant in the baPo area that led to the tragic events of August and September

2012 when close to fifty people were killed, most of them mineworkers. These events are etched indelibly on the public mind and caused an international outcry and considerable self-reflection in all sectors of South African society. Although this book is not directly about the revolutions (economic, social and environmental) that have accompanied the mining history of the bushveld, it does provide a context and background to some of the significant moments of the region's past that gave rise to these chaotic conditions. Moreover, labour-related issues are not the only source of socioeconomic discontent and division among the region's inhabitants. They have also been afflicted by deep-rooted forms of ethnic contestation.

A second image which the bushveld evokes today is that of its well-known game reserves. The two that stand out are the Pilanesberg and Madikwe reserves, but in recent years there has been a proliferation of smaller game farms in the area. Linked to the Pilanesberg reserve is the Sun City resort, a controversial island of opulence in what has been, despite the advent of mining, a very impoverished rural district. The development of these aspects of the tourism industry affected the surrounding communities in several ways – for example, it led to land alienation, and while work opportunities were created they were accompanied by exploitation and social dislocation.

A further outstanding characteristic of the region as a whole is its palaeontology which points to humankind's evolutionary past. The famous Taung child fossil was discovered in 1924 by Raymond Dart and the rich archaeology of the early baTswana has been excavated and written about by several of southern Africa's leading archaeologists.[10] Many important sites dot the countryside and provide evidence for its settlement by the forefathers of the African communities who still inhabit it. The sites range from small outposts, probably cattle kraals, to large towns or mega-sites inhabited by up to 20 000 people. The most spectacular were at Kaditshwene, north of modern Zeerust, and at Dithakong, the capitals respectively of the baHurutshe and baTlhaping in the early decades of the nineteenth century, both visited by European observers. More recently, large sites have been discovered at Molokwane and Marathodi, providing evidence of large-scale cattle keeping, extensive trade networks, an understanding of stone wall construction and the centralisation of power and wealth in the hands of a chiefly elite.[11] Many of the sites are within a day or two's walking distance apart, suggesting close political and trading relations between the groups who lived there. Although this volume's narrative begins well after these developments, the rich archeological evidence illustrates the longevity of the people in the bushveld/thornveld and their expertise as cattle keepers, and fashioners of iron implements and other trade items.

The archaeological evidence pointing to the origins and early settlement of Tswana-Sotho groups is confirmed by quite a rich body of oral traditions (pointing to when people have ruled and when specific events are said to have occurred) which trace the Tswana ruling lineages to as far back as the thirteenth and four-teenth centuries AD. The archaeology and the oral records prove conclusively that the pre-colonial baTswana did not live in conditions of tribal primitiveness or isolation and that they were thus capable of responding to, and engaging with, the new forces that swept across the western highveld from the mid to late-1830s.

Lastly, another view of the bushveld in particular is presented through the novels of Herman Charles Bosman written in the 1920s and based on his time in the Marico or Madikwe district. Particularly (but not solely) for generations of white South Africans for whom Bosman was prescribed reading at schools during the last fifty-odd years of the twentieth century, it is represented as a charming backwater of changeless quietude. Certainly, Bosman was too astute an observer not to ironically allude to the prejudices, hypocrisy and contradictions of Afri-kaner society in the Marico, but the overriding impression one gains is of a Boer society which first tamed and then laid claim to the region. In fact, it was a much more contested terrain than that, and the African societies of the bushveld exer-cised more independence than is presented in Bosman's novels.

The window through which all of these features of the North West Province can to varying degrees be viewed are the predominantly Tswana *merafe* or chief-doms: the baHurutshe, the baKgatla ba Kgafela, the baFokeng, the baKwena, the baKubung ba Rantheo and ba Monnakgotla, the baRolong, the baTlhaping and a number of smaller or related offshoots of these communities. Scholars examin-ing African settlement and social organisation in southern Africa have in recent times questioned the usefulness and precision of describing African societies in terms of 'tribes' or ethnic groups. The membership of these *merafe* was not fixed, and they were constantly being reshaped by newcomers or a changing of names. Nor can we simply assert that people a few centuries ago might even have called themselves the baRolong or baHurutshe. Despite this, we have adopted the 'eth-nic formulation' while recogising its limitations.

TRIBES AND ETHNICITY

Recent reformulations of the nature and meaning of 'tribes' among the baTswana (and in South Africa generally) have stressed their fluid, fluctuating and multiethnic character. Ethnicity is viewed therefore as a form of false consciousness, one that was imposed on African society by settler and colonial societies anxious to divide African people into recognisable and delineated 'tribes'. In 2010, Paul Landau extended this hypothesis by suggesting that there were other equally binding and durable forms of association and mobilisation that characterised African political organisation before their history was recorded and before 'tribes' or chiefdoms emerged as the key form of affiliation among Africans.[1] He rejected what he called the 'fog' of tribalism and the institution of chieftainship as colonial constructions.

The concept of 'tribalism' is open to manipulation, and has been recognised as such for some time. 'Tribes' are not primordial, and their size and composition changed over time owing to particular historical circumstances – yet they clearly had resonance and meaning for large numbers of the African population, and there is an equally cogent and countervailing view that ethnic groups are very real. As the late Nigerian political scientist Claude Ake, has pointed out:

> Apart from the question of its historicity, the logic for the argument for the non-existence of ethnic groups is flawed. Ethnic groups are no less real for existing intermittently, for having fluid boundaries, for having subjective or even arbitrary standards of membership, for opportunistic use of tradition … They are real if they are actual people who are united in consciousness of their common ethnic identity … ethnicity is not a fossilised determination but a living presence produced and even driven by material and historical forces.[2]

In a recent study of political responses to colonialism in south-western Zimbabwe, Enocent Msindo convincingly shows how Kalanga ethnic identity 'was not a creation of the colonial state and did not need to be', nor did these communities 'require colonial control to reinforce notions of community and identity'.[3] Moreover, Landau's significant initiators and actors of 'overlapping movements and … authority-building practices'[4] remain just as opaque as his 'fog of tribal peculiarism'. Thus we get a bewildering host of terms: 'princes', 'lords', 'fighters', 'patriarchs', who were members of 'rural hegemonies', 'admixtures, amalgamations', 'splittages' and 'regional patrol circuits'. Traditional leadership structures and figures are shifted to the margins and centralisation and state-building by powerful figures get relegated to a lesser category of mobilisation. We have decided to retain the term in its now common form of chiefdom or *morafe* in Setswana to discuss these bushveld communities, while recognising that they were not bounded and impermeable. In addition, from the mid-nineteenth century these communities occupied largely fixed boundaries.

NOTES

1   P. S. Landau, *Popular Politics in the History of South Africa, 1400 1948* (Cambridge: Cambridge University Press, 2010).
2   C. Ake, 'What is the Problem of Ethnicity in Africa?' *Transformation*, 22 (1993), p. 1.
3   E. Msindo, 'Social and Political Responses to Colonialism on the Margins: Community, Chieftaincy and Ethnicity in Bulilima-Mangwe, Zimbabwe, 1890-1930', in Limb, Etherington and Midgley (eds), '*Grappling with the Beast*', p. 155.
4   Landau, *Popular Politics*, p. 246.

This approach is largely a matter of convenience, for the sources – colonial, missionary or other – perceived and wrote about them in this way. In addition, the 'chief' became a lens through which a wider community was refracted.

From early in the nineteenth century, the baTswana in South Africa were influenced by Christian missionaries, in particular those of the London Missionary Society (LMS) and the Wesleyan Missionary Societies. In the western bushveld many became Lutherans under the direction of the Hermannsburg Mission Society (HMS), for whom land and agricultural production were essential material adjuncts to the evangelising mission. Land acquisition, with missionary encouragement and legal abetment, provided the cornerstone for economic security; it enabled many peasant producers and reserve dwellers to continue enjoying a relatively independent livelihood beyond the period when most other Africans in the country had joined the ranks of migrants or urbanised workers. Later (owing to mineral discoveries) ownership of land proved to be a windfall.

The LMS, who sought to convert the Southern baTswana to Christianity, placed an emphasis on education and the cultivation of European mores and cultural norms. Although it was often modified and adapted to suit their circumstances and practices, Christianity fundamentally challenged the ethical values and belief systems of the baRolong and baTlhaping, for example. People who became Christians often considered themselves to be a 'respectable' elite which was out of step with many aspects of traditional society. This changing identity was something many baTswana had to grapple with as they engaged with a new and 'modern' world.

We shed light on a number of themes that typify the history of the African people of southern Africa – migration, settlement, economic diversification, state formation, missionisation, colonisation, peasantisation, labour migration, accommodation and resistance. However, the book also guides readers to the more unusual aspects of the region's past and its significance for the present, especially as far as the black population is concerned.

Issues and themes

Between about 1820 and the late 1830s, the *difaqane*, a period of conflict and rapid transformation that involved many African societies, was ushered in across southern Africa. Almost immediately afterwards, white trekkers established a state on the western highveld. Significant new leaders emerged after these turbulent years. Chapter One recounts the life of a hitherto neglected mid-nineteenth century leader, Moiloa II of the baHurutshe. Like a number of his contemporaries such as Montshiwa of the Ratshidi baRolong and Mokgatle Thethe of the baFokeng, Moiloa showed the ability, coupled probably with some good fortune, to re-build his community from the 1840s through to almost the mid-1870s. It meant dealing with a more complex set of circumstances than he had experienced before. His ability to strike up a good relationship with powerful Boer personalities, in particular Jan Viljoen in the Marico district, was critical to the reconstruction of Hurutshe society.

The North West Province was one of the most hotly contested battlegrounds of the South African War. Our book adds to the growing historiography on the role of black communities in that war by providing, in Chapter Two, a holistic sweep of the encounters between the Boers and the local African populations. The conflict played itself out on different levels – on the battlefield, on Boer farms, and in Tswana villages and towns. We also discuss its aftermath, showing that its repercussions and impact were profound for the African population; albeit for a short moment, factions within African society tried to reverse the inexorable tide of segregation and dispossession that had swept over them by seizing Boer assets.

The next three chapters examine aspects of economic and social change among the province's black population in the first half of the twentieth century. As alluded to above, land acquisition enabled African communities in the western Transvaal to keep livestock and maintain a reasonable level of agricultural production. The bushveld inhabitants and their rulers were especially concerned with buying or gaining access to farms, acquired either with missionary assistance or in defiance or ignorance of legislation designed to prevent Africans from purchasing land in the Transvaal.

In the Bechuanaland reserves, from Mafikeng south to Vryburg and Kuruman, owing to circumstances relating to Britsh colonial expansion the African population was allocated inalienable land, distinct from the government or crown land from which Africans could be removed. This provided a measure of security from potential land-grabbers of various kinds.

The need to obtain and retain a hold on land is a recurring motif in the history of the baTswana of the western Transvaal. When the 1913 Natives Land Act was passed (demarcating the limits of African land holdings) just about every western Transvaal chief enthusiastically supported the South African Natives National Congress (SANNC, later African National Congress), hoping it could reverse the deleterious consequences of the legislation. Their enthusiasm waned in most cases when it became apparent that the ANC was incapable of doing so. On the other hand, a negative and potentially destructive consequence of land acquisition was that it created discord between rulers and subjects over access to and control of land and the material resources it provided.

Thus in Chapter Three we examine the internal crises of authority and chaos that afflicted the baFokeng, the baKwena ba Mogopa, the baKgatla ba Kgafela and the baKubung bushveld *merafe* from about 1902 to the middle of that century. This instability was caused primarily by questions of control over material resources. Although they led to social stress and conflict, the basic symmetry of traditional society was not significantly eroded by these conflicts, but simultaneously individuals began to see themselves as independent from the framework of traditional society. What is important about these crises is that they reoccur in an even more contested form after the platinum revolution.

Chapter Four examines the fortunes of the baRolong and, to a lesser extent, the baTlhaping, who inhabited the reserves set aside for them in the late nineteenth century. Confined to an area smaller than they had occupied before colonisation, and restricted by colonial legislation, these rural communities endured considerable hardships but successfully negotiated the transitions of the first half of the century. The nature of Rolong politics was historically rooted and complex, but most of the significant conclusions to this past were resolved in the first half of the nineteenth century. Chiefly political struggles for local control were commonplace, and were closely related to resource control in a difficult environment, especially where the jurisdiction of traditional authorities was defined and limited. The growth of a progressive rural elite and an administration generally supportive of innovation, education and 'advanced' farming methods combined to keep the reserves sustainable and limit their incorporation into the political and economic framework of the colonial state.

Chapter Five has a different focus in that it consists of a re-appraisal of an event known as the 1957/58 Bahurutshe Revolt. We give an account of what actually transpired and its noteworthy features, and we then analyse the reasons for the revolt which was led by the *kgosi*, Abram Moiloa, and was strongly supported by the women of what was called Moiloa's Reserve, north of Zeerust. The reason for the resistance is considered to have been the order by the South African government that women should carry passes, as African men had been forced to do some years earlier. But this was the straw that broke the camel's back – yet another incident in a continuum leading to the collapse of the sustainability of Moiloa's Reserve. We profile the history of the reserve from the early nineteenth century in order to show how it was these longer-term hardships that best explain the depth of anger and frustration of the women in particular in the reserve and their determination to resist the carrying of passes.

From the 1950s, the apartheid system in South Africa created a new set of circumstances for most of the inhabitants of the region. They were herded into the Bophuthatswana homeland, and faced new challenges, no less demanding or difficult than before. In an attempt to gain full political and economic control, especially of the platinum mines, Lucas Mangope and the Bophuthatswana government intervened in the affairs of the baTswana in the bushveld and this led to a spate of conflicts and resistance, with dire consequences. In Chapter Six we describe how these materialised and their resolution, which included some of the most vicious litigation in the history of mining law between the baFokeng on one hand, and the big mining companies and Mangope on the other. Significantly, changing concepts of ethnicity define the Mangope years, at a time when the 'bantustan' concept allowed for flexible interpretations of ethnicity.

Finally, we look at the conditions prevailing in South Africa's 'Platinum Belt' in modern times and provide observations as to how it has been radically transformed in recent years, although the transformation was by no means an even one. This chapter summarises the impact of the expansion of the platinum mining sector and the massive windfall that accrued from it, on nearly all the baTswana in the Rustenburg region. In important ways it has been beneficial: for example it has enabled the baFokeng, the first and major beneficiaries of mining on their land, to corporatise their affairs and, paradoxically, to realign themselves as a 'nation' headed by a traditional ruler (termed their king) and governed by a Royal baFokeng Administration. But the massive profits from mining, coupled with the financial deals concluded with the local *dikgosi*, under the terms of black economic empowerment (BEE), have led to the rise of mineral-based ethnic assertion – and this in turn has created the possibilities for huge personal accumulation and the eruption of social turmoil on a scale not seen before. By 2013 there was hardly one African ethnic community in the bushveld that was not embroiled in a malevolent contest for the various earnings from mining on their land, or for control or ownership of the land itself. These events unfolded in the context of the 'new' South Africa, where rights were considered inalienable, and where the state, capital, labour and 'traditional' communities contended for mining revenues.

Finally, another face of the bushveld region in particular is represented by its national game reserves and the vast entertainment complex of Sun City. We look at the consequences of this for the local communities. While profitable, certainly in the long-run, to significant sectors of the population, it served only to further marginalise and impoverish other segments of the rural population, some of whom were removed from their homes to make way for the parks.

ENDNOTES

1   S.M. Molema, *Montshiwa, 1815-1896: Barolong Chief and Patriot* (Cape Town: Struik, 1966) .

2   K. Shillington, *The Colonisation of the Southern Tswana, 1870-1900,* (Johannesburg: Ravan, 1985); K. Shillington, *Luka Jantjie, Resistance Hero of the South African Frontier,* (Johannesburg, London and New York: Wits University Press, Aldridge Press and Palgrave Macmillan, 2011).

3   A. Delegorgue, *Travels in South Africa*, 2 vols. (Scottsville: Natal University Press, 1997), pp. 237-244.Translated by F. de B. Webb.

4   J. Campbell, *Travels in South Africa … Narrative of a Second Journey, 1820*, (London, 1822), p. 220.

5   W.C. Harris, *The Wild Sports of Southern Africa* (London, 1861), p. 195.

6   A.A.l. Agar-Hamilton, *The Road to the North, South Africa, 1852-1886* (London, 1937), p.1.

7   The phrase is Martin Legassick's from 'The Griqua, the Sotho-Tswana and the Missionaries, 1780-1840, The Politics of a Frontier Zone', Ph.D thesis, University of California, Los Angeles, 1969, p. 380. Landau prefers the term 'metis'.

8   See A. Manson 'The Hurutshe in the Marico District of the Transvaal, 1848-1914', PhD thesis University of Cape Town, 1990, p.31, citing N. Parsons, 'Khama III, the Bamangwato and the British: with special reference to 1895-1923', Ph.D thesis, University of Edinburgh, 1973, p. 77.

9   N. Parsons, 'Khama III, the Bamangwato and the British', p.77.

10 See for example, R. Mason, *Prehistory of the Transvaal: A Record of Human Activity* (Johannesburg: Witwatersrand University Press,1969); T. Huffman *Handbook to the Iron Age: The Archaeology of Pre-Colonial Farming Societies of Southern Africa* (Scottsville: University of KwaZulu-Natal Press, 2007); T. Maggs and G. Whitelaw, 'A review of recent archaeological research on food-producing communities in southern Africa,' *Journal of African History*, 32 (1991); J.C. Pistorius, *Molokwane, An Iron Age Bakwena Village: Early Settlement in the Western Transvaal* (Johannesburg: Perskor, 1992) especially pp. 20-32, 38-46; J.C. Boeyens, 'The Late Iron Age sequence in the Marico and early Tswana history, *South African Archaeological Bulletin*, 58, 178 (2003) and S. Hall, 'The Late Precolonial Tswana in the Rustenburg District', in N. Swanepoel, A. Esterhuysen and P. Bonner, (eds) *Five Hundred Years, Rediscovered* (Johannesburg: Wits University Press, 2008).

11 Hall, 'The Late Precolonial Tswana'.

CHAPTER 1

'The dog of the Boers'?
Moiloa II of the baHurutshe

INTRODUCTION

We know a lot more nowadays about important chiefly personalities in the history of the baTswana in South Africa. Luka Jantjie – despite initially placing faith
in British justice and European religion – found his people subject to colonial
laws, and his country overrun by white colonists. In order to defend shrinking
independence and constant land alienation he resorted to a final desperate act of
military defiance in 1897.[1] Montshiwa, similarly, fought for a quarter of a century
to protect his territory in the Molopo region from the ambitions of white mercenaries, as did Mankurwane of the baTlhaping further south near Vryburg.[2] Both
finally opted for British protection, despite the significant loss of autonomy that
this meant. The career of Mokgatle Thethe of the baFokeng near Rustenburg has
now also been revealed. As the 'founding father' of the baFokeng, he formed a
close association with Paul Kruger, later president of the South African Republic.
This enabled him to buy a much needed measure of independence and to embark
on a programme of extensive land acquisition which was later to form the basis
for baFokeng material security and, later, mineral wealth.[3] To preserve their land,
their independence and ethnic unity, these men resorted to tactics ranging from
outright resistance to accommodation with the colonising forces – but they lived
in difficult and complex times, and to view them as mere collaborators (as has
been the case with Mokgatle) would be an oversimplification.

Our focus, though, is on a *kgosi* who has not received full recognition for the role he played in reconstituting and laying the foundations for the continued security of his society. In many respects Moiloa's career mirrors that of other nineteenth-century Tswana leaders. It also reflects some of the key features of the experiences of African communities from the mid-1840s to the turn of the century. Reconstructing his life means also examining crucial external forces and institutions such as the missionaries, the local Boers who had moved onto the western highveld from 1838, state officials of the South African Republic, and Tswana neighbours.

Introduction

In 1800 the baHurutshe lived about twenty kilometres north of the present-day town of Zeerust in the wider Marico district. Archaeological evidence reveals that they had been in this locality for close to half a century. How far back we can trace the baHurutshe as an identifiable community calling themselves by that name is a moot point; probably, as with all Tswana *merafe*, there had been almost constant fission, breakaways, regroupings and new arrivals so that the composition of the baHurutshe was constantly changing. They certainly were one of the larger factions of what has been termed 'lineage-clusters' (people related by common descent) in the western bushveld. For periods before 1800 they enjoyed privileged status among the local residents, though this pre-eminence was probably not continuous.[4] What is more certain is that they were engulfed in a series of localised intra-baTswana conflicts from about 1790 to 1820, when they were attacked by new raiders from the south.[5]

THE 'TSWANA WARS' AND THE *DIFAQANE*

The *difaqane* has been translated variously as 'the crushing' and 'the time of troubles'. Prior to the 1970s, it was generally thought that these changes derived from the growth of the Zulu kingdom under Shaka in the south-east, and that the changes occurred from the end of the eighteenth century up to the 1820s. Such views were later challenged and, as a consequence, modified. The impact of the Zulu on their neighbours has been questioned and the geographic focus of the process has been widened to include the interior of South Africa (in particular the baTswana of the western highveld), and the beginning of the *mfecane* has been extended back from about the 1790s to the mid-eighteenth century.[1]

Earlier accounts and versions focused on events in the territory from the Thukela River to Delagoa Bay in the south-east. It was noted that from about 1780 certain

African chiefdoms expanded in size and power, whereas weaker ones were displaced or incorporated. In response, most chiefdoms looked to bolster their military capabilities, and by 1818 two dominant forces had emerged: the Ndwandwe under Zwide in northern Zululand and the Mthethwa under Dingiswayo further south. Among Dingiswayo's allies were the Zulu, whose leader, Shaka, had been installed by Dingiswayo. The Zulu subsequently established a powerful state. To avoid the growing conflict, several local leaders took their followers out of the Zululand region, across the Drakensberg and north to the present-day Swaziland/Delagoa Bay region, and west onto the central highveld. From the perspective of the baTswana on the western highveld, the most important of these leaders was Mzilikazi of the Khumalo, who led his followers out of present-day KwaZulu-Natal because he may have crossed swords with Shaka (as old-style historians have suggested) or because he simply sought a more peaceful and secure home for his followers.

The next obvious question asked by historians was why there should have been such a relatively sudden spate of upheavals and political realignments. Various ideas have been propounded: overpopulation, the effect of climate changes caused by drought, environmental degradation and competition for good pastures were among those put forward in the 1960s. In the next decade, the effect of the entry of new trade goods and the competition created for control of this trade, coupled with European demand for ivory, gold and slaves, were propagated as reasons for the sudden shift to more extreme military measures and organisation. In the 1980s, a group of historians led by Julian Cobbing, building on earlier ideas of Martin Legassick, shifted the debate away from the Zulu and other African communities to advance the idea that the 'time of troubles' was caused by expeditions and raiding parties who wanted to seize labour and slaves from Africans living in the interior. These were inspired, organised and conducted by whites (or colonial surrogates such as the Griqua and Kora) living at the Cape or in Portuguese Mozambique. While the evidence for such activities has not been consistently convincing, these scholars have reminded us that there were other actors apart from the Nguni states that were exerting significant influence on the affairs of South Africa's interior regions several decades before formal colonisation.[2]

The baTswana were broadly affected by these developments at least two decades before the first 'raiders' arrived from Nguni land. Oral traditions point to 'general restlessness and instability'[3] (sometimes called the 'Tswana wars') among the various *merafe* from the late eighteenth century. These were caused by pressure on good pastures, raiding for cattle and the seizure of women as captives, and competition for control of trade items, especially ivory which was in great demand in the East.

NOTES

1   For more detailed information on the 'Mfecane debate' see the contributions in C. Hamilton (ed.), *The Mfecane Aftermath: Reconstructive Debates in Southern African History*, (Johannesburg and Scottsville: Witwatersrand University Press, Natal University Press, 1995).

2   For these developments see in particular J. Cobbing, 'The Mfecane as alibi: Thoughts on Dithakong and Mbolompo', *Journal of African History,* 18 (1977); and most chapters in Hamilton (ed.), *The Mfecane Aftermath,* and more recently N. Etherington, *The Great Treks: The Transformation of Southern Africa 1815-1854,* (Harlow, 2000).

3   N. Parsons, 'Prelude to the Difaqane in the Interior of South Africa, c. 1600-1822, in Hamilton (ed.) *The Mfecane Aftermath,* p. 323.

The career of Moiloa II

Moiloa was born in about 1796 and would have been a young man when these conflicts broke out. By 1821 his father, Diutlwileng, had been killed, probably by raiders under Sebetwane of the Patsa-Fokeng. When the LMS missionary John Campbell visited the Hurutshe capital at Kaditshwene, he reported that a 'gloomy spiritlessness' pervaded the townspeople and that Hurutshe regiments *(mephato)* were patrolling the perimetres of Kaditshwene.[6] Moiloa's uncle, Mokgatlhe, was acting as regent, but Campbell remarked on Moiloa's popularity, and predicted (correctly) that he might 'wrest control from [Mokgatlhe's] hands'.[7] Between April and September 1823, further raids led to the abandonment of the town and the baHurutshe fled westwards to the hills of Mosega. However, a new threat was emerging. Mzilikazi's amaNdebele were now taking total control of the western highveld and expanding ever westwards. This forced Moiloa and Mokgathle to flee further southwards where they were found in a state of near destitution by a group of French missionaries of the Paris Evangelical Missionary Society, one of whom recorded that the baHurutshe, 'in the interval of a day, found themselves reduced to a diet of meagre roots'.[8] The French missionaries put their number at a paltry 700 to 800 souls, compared to the 'thousands' they had encountered at Mosega, just west of modern-day Zeerust.

Though invited by the missionaries to join them, Moiloa and Mokgatlhe chose to move, in 1834, to a place called Modimong on the Harts River. Here they attached themselves to the Kora, an independent Khoekhoe community under David Mossweu. The baHurutshe under Mokgatlhe and Moiloa then joined an aggressive Kora/Griqua alliance to try and force Mzilikazi out of the Madikwe region; thus in 1834 a commando sent against Mzilikazi contained within its ranks a contingent of Hurutshe men under Moiloa, who returned with a few hundred cattle.[9] This alliance was strengthened by the arrival of a trekker party under Andries Potgieter, who had left the Cape Colony in 1836. In January 1837 the two groups joined forces to drive the amaNdebele out of their military fortresses. Once the amaNdebele had left the western highveld, Moiloa was quick to appreciate the power of the trekkers, and personally visited Potgieter soon after their arrival to ask if he could resettle in the former Hurutshe homeland, to which Potgieter agreed – however, over a decade was to pass before the move was finally made.

There were several reasons for this lengthy delay. First, there was a possibility that the amaNdebele might return, and there was no way of knowing if the trekkers would settle once and for all in the western highveld or if they would continue to hold power in the region.

Second, there was a power struggle among the baHurutshe over the succession. Moiloa and Motlaadile were the sons of the deceased *kgosi* Sebogodi. Their older brother Menwe had died before their father, but Mokgatlhe their uncle then ruled the chiefdom as they were both minors at the time of Sebogodi's death. Mokgatlhe had in turn married Menwe's main wife and had 'raised up seed' on behalf of his dead nephew and fathered a son named Lentswe. Thus there was a three-way rivalry for power between Moiloa, Motlaadile and Lentswe and this jockeying for the chieftainship seemed to preoccupy Hurutshe politics for several years. The alliance between Moiloa and his uncle did not fall apart, despite Mokgathle's claiming of the right to leadership for his son Lentswe. Of the three men, however, Moiloa was the most politically astute, and grasped the realities of the time better than the others. He enlisted the aid of the militarised Griqua under their leader (or *kaptyn*) Waterboer, and requested his assistance in the relocation of the baHurutshe to the Madikwe region. Not only could the Griqua offer military support in times of need, but they could also provide other skills and services (for example literacy) which might stand the baHurutshe in good stead. Moiloa also approached Walter Inglis of the LMS in Griquatown in about 1838, and by all accounts the idea of a mission to the baHurutshe was discussed during this visit.

The LMS was in fact already preparing to receive the baHurutshe in the Madikwe district. In 1842, David Livingstone, famous later for his explorations in southern Africa, had established a station among the baKgatla ba Mmanaana at a place called Mabotsa, near present-day Gopane, and it was proposed that the baHurutshe be settled nearby, at Mangelo River, some five kilometres from the old town of Kaditshwene. Between April 1847 and September 1848 the baHurutshe under Mokgatlhe and Moiloa moved to this vicinity. They were finally home again. Their missionary, Walter Inglis, was delighted. He wrote to his superiors:

> I am happy to say Moiloa, my old chief, has joined me with his people … He was with me last Sabbath. It was by far the largest meeting [of the baHurutshe] I have ever had. It had been a great misfortune to me that the baHurutshe did not come on all at once. The whole question is an involved map of native politics. Now Moiloa has come the scattered villages will be gathered together. Moiloa … will add weight and respectability to the mission.[10]

By this time Mokgatlhe had allegedly fallen out with his son Lentswe, who went to join Montshiwa's baRolong; and Moiloa, who was now in his mid-forties, had taken effective control of the chiefdom.[11] He was to enjoy a long and important reign.

Matters seem to have gone well for the first year. The Boers in the Transvaal were thinly spread and concentrated nearer to the settlement at Potchefstroom. Andries Potgieter had moved with his followers to the eastern Transvaal in 1845, and the other trekker parties were preoccupied with attempts to expel the British from the Orange River Sovereignty; however, after the defeat of the Boers by the British at the battle of Boomplaats in August 1848, they fled over the Vaal River and by early 1849 were beginning to encroach onto the land of the baHurutshe. Trouble was looming.

The Marico trekkers shared similar aims to the African people in that frontier zone. Both white and black societies wanted to settle on the land and gain control over land and labour, and both had just undergone periods of extreme disruption. One of the prominent trekkers, destined to play a significant role on the frontier, was Jan Viljoen. As a reward for serving under the forces of Pretorius in the battle of Boomplaats in 1848, he and a number of other trekkers received farms close to the Klein Marico River, some of them as large as 30 000 morgen. The trekkers who arrived some years later and who had not been rewarded with land grants were those who sought to gain land at the expense of local African chiefdoms such as the baHurutshe.

Thus the first problem confronting Moiloa was the encroachment of the Boers onto land that the baHurutshe considered theirs. In early June 1849 Rodger Edwards (who had joined Inglis) reported that a number of Boers 'had located at Mosega on the streams in the vicinity about twenty miles distant. Their future progress will be northwards.'[12] Anticipating trouble, Edwards met with Potgi-eter and was assured by him that both the baHurutshe and the missionaries had nothing to worry about. But relations remained strained. By September, Edwards reported again that the trekkers were 'determined to occupy every available foun-tain and are resolved upon making chiefs and the people bow to their rule'.[13] At the Manegelo River, Moiloa had received an order from the Boer authorities to provide labourers, and thought it more prudent to move away from the advancing Boers to Dinokana (place of many streams), about twelve kilometres away from them. With him were 1 500 followers and about fifty Griqua converts who had been 'given' to him by Andries Waterboer. Dinokana was a good site and remained the centre of Hurutshe settlement. The new mission station was called Mathebe and here Moiloa began the long and laborious task of reconstructing his *morafe*.[14]

There were several advantages favouring Moiloa's baHurutshe. The first was that they had been promised land by Potgieter, and this location formed the basis of what became known later as the Hurutshe Reserve or Moiloa's Reserve. The grant was in return for Moiloa's assistance in providing men for the campaign against Mzilikazi in 1837. In 1865 the Volksraad (parliament) passed a resolution defining the reserve and adding to its original size. Later it was estimated at 125 584 morgen – the largest tract of land set aside for African occupation in the western Transvaal (the creation of the Hurutshe Reserve was unusual, for most Africans in the former western Transvaal at this time lived on privately owned land). The second advantage was that the number of Moiloa's followers increased as he was joined by many of the splinter groups that had gone their own way from 1823. These included two of his brothers, Motlaadile and Pule (though the latter's relationship to Moiloa is not absolutely clear) and his nephew Sethunya. In August 1850, Inglis reported the arrival of a party of baHurutshe in Dinokana from a 'town east of the Limpopo' [river].[15] By the early 1860s the population numbered about 8 000. They were probably not all 'pure' baHurutshe, but included the Griqua converts and a number of strangers (subordinates or auxiliaries) encountered during the years of exile and wandering south of the Molopo River.[16]

An important event occurred during these years. Lentswe was killed in a skirmish with some Boers near Lotlakane, some sixty kilometres from Dinokana, in the territory of the raPulana baRolong. Lentswe's son, by rights, had a prior claim to the chieftainship, an issue that had complex and divisive implications later on. But his death meant that for the time being at least there was no effective opposition to Moiloa, as Lentswe's son Gopane was only about six years old. It was therefore futile for any opponents of Moiloa to try to manipulate succession laws to elevate a possible rival to power. Moiloa was the *kgosi* accepted by those who saw him as the rightful successor to Diuwitleng or by those who claimed Gopane as the rightful chief.

Thus the baHurutshe not only had access to sufficient land and an abundance of spring water at Dinokana; they also had an unchallenged leader in Moiloa who was backed by a majority of the *morafe*. His main problem, however, was that the trekkers resorted to frequent and random attacks. In one sense, these attacks were an expression of their own weakness and inability to control the African population through laws and regulations (the Boers were at this time still battling to form an effective state on the highveld). Furthermore, the LMS missionaries soon proved to be more of a burden than anything else for the baHurutshe. Their presence antagonised the Boers and they failed to prevent aggression against their followers.

What lay at the root of these attacks was the system of 'apprenticeship' introduced by the trekkers, something that deeply affected nearly all African people living in the bushveld during these years. Soon after they arrived in the Transvaal, Boer commandos began periodic raids on weaker and less organised African communities with the intention of capturing their children in order to use them as 'bonded labourers'.[17] As the Reverend Freeman, on a visit in 1849 to the LMS stations among the baHurutshe and baKgatla ba Mmanaana at Mabotsa and Mathebe, graphically recorded:

> … a party of armed Boers came and demanded orphans who might be there … after much altercation and the steady refusal of the chief to give up the orphans, the Boers demanded the children of the people. The Boers began to seize them and put them into wagons; the men interfered; the Boers fired, and in the result most of the men were killed defending their families and the wagons were loaded with the children and driven off as booty.[18]

Sometimes the children were demanded as tribute, or were traded, or secured through exchange. Such captive children, known in Dutch parlance as *inboekelinge* (registerees), were 'booked in', as 'orphans', and indentured to their masters.[19] Rustenburg commandos, for example, raided African groups in the far northern Transvaal – indeed, Rustenburg has been described as 'a slave trading centre with its own resident dealer' (the 'dealer' referred to was none other than Paul Kruger, the future president of the South African Republic, later the Transvaal).[20] The young captives were shared among the Boer commandos and brought up on their farms.

As servants on the Boer farms, the *inboekelinge* were trained in a variety of skills: stonecutting and building, brick making, cookery, veterinary and folk medicine, wagon repair, hunting, gun maintenance, making cheese and plough farming. In time they adopted the customs, norms and language of the trekkers and became known as *oorlams*. Perhaps the most important use to which the *oorlams* were put was that they were trusted with firearms, and became expert hunters who accompanied the Boers on hunting and trading expeditions. In using *inboekelinge* labour, the Boers were in fact continuing an old tradition they had brought with them from the Cape where indentured Khoekhoe and coloured labourers were used for a variety of unpaid domestic and military functions.[21] The *inboekelinge* males were supposedly manumitted (freed) at the age of twenty-one and the females at twenty-five. Many of them, however, were alienated from African society and chose to remain within Boer households – others such as the residents of the Bethlehem location in Rustenburg and a group of *oorlams* who occupied a farm (Welgeval) in the Pilanesberg crater, lived in distinct *oorlam* communities.[22]

Several Setswana-speaking chiefdoms tried to resist the hardship imposed on them by the trekkers, and this frequently led to direct armed confrontation. The best-known incident is that of the attack on the baKwena and Livingstone at Dimawe in Botswana, an event that had direct implications for Moiloa's baHurutshe living close by. The confrontation, however, had its origins in the Transvaal when Mosielele, the *kgosi* of the baKgatla ba Mmanaana at Mabotsa, and neighbour to Moiloa, decided to flee the Transvaal and seek refuge with Setshele's baKwena after repeated demands to provide labour which he had ignored. The Marico Boers decided it was time to punish such 'impertinence'. Commandant A Stander advocated sending a commando against Setshele, but Jan Viljoen initially opposed the idea. The general Boer determination to punish the baKwena was, however, unstoppable and a commando attacked Setshele and Mosielele in 1852. The attack was a disaster for Africans living alongside the western Transvaal's border. The baHurutshe under Mangope living at Borutwe were forced to assist the commando, as was a contingent of Moiloa's men who headed the commando sent to torch Setshele's town.

After this incident Moiloa decided to abandon Dinokana. He sought protection first among Setshele's baKwena themselves and then with the baNgwaketse under Senthufe, further west in today's southern Botswana. This was in itself an indication of the 'diplomatic' contacts Moiloa had forged with the independent western baTswana. Viljoen encountered him in January 1853 at Chonwane, Setshele's former capital, about eighty kilometres north of Mathebe, and asked him why he had fled during the raid, to which he replied that other *dikgosi* around him had also fled, and that furthermore he 'was looked upon as a traitor because he lived among whites'.[23] Moiloa nevertheless affirmed that he wanted peace, and Viljoen asked him to return to the Transvaal. The meeting ended with an interesting exchange: Viljoen insisted that Moiloa observe the labour regulations (that *dikgosi* render up men for Boer work parties) to which he reportedly replied, 'No, don't ever ask me for people, I have too much to do myself, but keep the road of peace open and I will see to it that people come to you as before.' By August, most of Moiloa's followers had returned to Dinokana, although he personally was still resident with Senthufe, and only returned after another appeal from Viljoen.

The attack on Dimawe and its consequences also signalled the end for the LMS in the Transvaal. The missionaries Edwards and Inglis remonstrated with the authorities of the South African Republic about the incident and Edwards criticised it in a local newspaper. The Boers were incensed and charged the missionaries with high treason (for allegedly supplying guns to the Africans). After rather comic opera court proceedings the two men were found guilty and expelled from the Transvaal.[24] This left Moiloa stripped of the potential backing of the LMS although, given the antagonistic relationship between the British missionaries and the trekkers, this may have been a blessing in disguise.

By this time the Setswana-speakers in the western Transvaal were beginning to understand that they were dealing with a society that was far more powerful than even the amaNdebele had been, and that it would be difficult to retain a hold on the crucial resources of land and labour, as well as to retain a semblance of political autonomy. Coupled with this was the clear military and technological superiority of the whites. This realisation led Moiloa to make the disconsolate observation in 1852 that he was nothing but a 'dog of the Boers'.[25]

But the trekkers had by no means created the conditions for exercising supreme control in the Marico district. In 1849 they convened a meeting at Deerdepoort to try and establish a government of unity for the Transvaal, but Potgieter stood aloof from the meeting. Provision was, however, made for a *Volksraad* (parliament), and some officials were appointed to key positions such as commandant-general to oversee African affairs. Despite this, the state was administratively weak and financially bankrupt and remained so even after the formation of the South African Republic in 1852. African leaders like Moiloa realised this and began to see that they could avoid the worst cases of Boer authority.

The Dimawe affair has been fully recorded, largely because their missionary David Livingstone was himself a victim and widely publicised it. But subsequent events in the bushveld are less well known. Once the commando had been disbanded and returned home laden with booty, the Marico Boers were at the mercy of reprisal attacks from the baKwena and the baRolong (who had also been attacked) who wanted to recapture their stolen cattle and children. Among those captured was Setshele's son Kgari, and the failure of the Boers to return him was a source of ongoing anger to the baKwena. During 1852 and 1853 three Boers were killed in minor skirmishes, the Marico farms were abandoned, and the Boers went into *laager*. In January 1853 James Chapman, the English explorer, hunter and trader, encountered 200 wagons headed for the security of Potchefstroom. Hunting was suspended and several Boer farms were looted. The attack, moreover, divided the Boers. Viljoen, representative of the 'hunters faction' was openly critical of Commandant Scholtz who had led the commando, accusing him of having caused the 'ruination of the inhabitants by his wanton proceedings'.[26] In fact the Marico Boers almost abandoned the district, a misfortune which befell other trekker frontier outposts in the Transvaal such as at Ohrigstad and Schoemansdal. It was left to Viljoen and his 'hunters faction' to negotiate peace terms with Setshele. Although the meeting was successful it took several years before the trekkers returned to their Marico farms.

It was not only the Boers who felt the insecurity of the Marico. Some of Mangope's baHurutshe at Borutwe (Mangope's Siding) fled the Transvaal and took refuge with Setshele. This kind of 'protest migration' was made possible by the proximity of the Transvaal border with the independent and frequently related Setswana-speaking chiefdoms living just across the line. Even after the formal restoration of peace, instability continued as a result of the murder of three Boers and retaliatory attacks on groups of baKwena living in the Transvaal; Viljoen established that the culprits were subjects of Makopane's Nzundza or Transvaal

Ndebele in the Waterberg district. The borderland between the Transvaal and the baKwena remained intermittently unsafe for all, although peace, largely as a result of Viljoen's efforts, was finally restored. For several decades, relations between the South African Republic and the independent baTswana *merafe* became what has been characterised as a 'diplomatic game' between the two leading figures, Setshele and Jan Viljoen, who has been portrayed as a peacemaker along the Marico frontier.[27] There certainly is good reason to include a third player in Moiloa, who assumed the role of intercessor between the two men.

The Dimawe affair and its consequences was revealing to the more capable African leaders in the region such as Moiloa. It became evident that the trekkers were unable to bring the western Tswana chiefdoms under their control (or to expand in their direction) and that they were thus not as all-powerful as they at first appeared. In addition, in order to keep the important 'hunters road' to the north open, the trekkers had to rely on certain African allies. Moiloa, one such ally, realised he could buy some breathing space by offering support to the Boers when convenient. This mutually dependent relationship bears a strong resemblance to that which developed elsewhere in the Transvaal between factions of trekker and African society.[28] Thus it was, for example, that negotiations between a number of African leaders and the Boers took place at Dinokana; that Moiloa acted as a messenger between them; and that he reported any incidents which might threaten the peace of the district. During the troubles between the South African Republic and Setshele's baKwena, Moiloa and his entire council held meetings with Viljoen on his farm at Vergenoeg, where Moiloa offered 'to make further enquiries into the rumours [of impending war] and to punish the guilty'.[29] The extent of the support Moiloa was prepared to give the authorities is suggested by a statement he made in 1856 to Pretorius that 'if a fly falls in the milk from my side I will take it out so you [Pretorius] can punish the culprit'.[30] By adopting a cooperative position, Moiloa was in fact making himself indispensable to the Boers, and was later able to make certain demands on them.

For nearly all the Setswana-speaking communities in the bushveld, missionaries became a necessary precondition for independent political and economic activity, and Moiloa consequently turned his thoughts to this soon after the expulsion of the LMS. Of course, it entailed making compromises. In return for their support, missionaries expected that African *dikgosi* would create the right circumstances for successful proselytising and conversion to Christianity. Most active of the missionary societies in the region was the Hermannsburg Mission Society (HMS). After the expulsion of the LMS from the Transvaal, the authorities sought missionaries who would encourage their converts not to challenge the supremacy of the Transvaal state. In 1858 the baKwena paramount Setshele had asked the South African Republic to assist him in finding missionaries for his followers in

Botswana and President MW Pretorius had no doubts about approaching the Hanoverian mission in Natal. 'In their schools,' he wrote, 'they concentrate on encouraging the barbarians to work and on giving them a sound conception of the secular order of affairs before instructing them in the divine.'[31] The HMS thus revealed little of the humanitarian zeal displayed by many of the LMS missionaries. As one of them explained to the Reverend John Mackenzie of the LMS, 'We Hermannsburgers are so deficient as politicians that we cannot dispute the supremacy of the South African Republic over the Bechuana tribes.'[32] Nevertheless, the HMS missions were generally established at the request of the African population, and by not antagonising the state authorities they were generally left undisturbed to serve their communities.

THE HERMANNSBURG MISSION SOCIETY

In addition to ministering to the baHurutshe and baFokeng, the HMS also worked among the baKwena-ba-Magopa at Bethanie, where the Reverend W Behrens settled in 1864. Before the turn of the century, the HMS also founded stations among the baPhalane in the Pilanesberg district (Kroondal), the baPhiring in Mabaalstad (Emmaus), at Pella among the baKwena ba Modimosana, and among the *oorlams* people in Rustenburg town. Thus, with the notable exception of the baKgatla-ba-Kgafela, the HMS had a monopoly over missionary work among the African population in the bushveld during the nineteenth century.

Some time near the end of 1858 Moiloa requested the HMS missionaries then with the Bakwena at Diteyane in Botswana to visit him, and three missionaries led by Reverend Ferdinand Zimmermann came over to Dinokana. They were quick to spot the good agricultural potential of Moiloa's location. Moiloa offered them a large site on which they could establish a mission. Zimmermann obtained the permission of the SAR authorities to work among the baHurutshe and soon after the establishment of the mission Zimmermann lived up to Moiloa's expectations of him as an intermediary with the SAR authorities by requesting guns and ammunition for Moiloa's people to hunt with and to protect their cattle from wild animals.[33]

The Tswana *merafe* of the north-west highveld generally welcomed missionaries for the range of services they could offer; they were useful, for example, as emissaries and diplomatic agents for the *dikgosi* or their representatives, and introduced their converts in particular to new economic methods and concepts – in particular that of land ownership. Their influence was not limited to the realm of the political economy however: they largely dismissed African cultural practices and beliefs and encouraged Africans to embrace an awareness of the ideas and cultural beliefs of Europeans. They therefore provided a window into the world and the consciousness of the *makgowa* (Europeans). The significant leaders, including Moiloa, initially resisted conversion to Christianity, probably because it would have risked alienating powerful traditionalists within their communities but, nevertheless, relations between him and the missionaries were cordial. Moiloa encouraged children to attend the mission school and was considered by Ferdinand Jensen, a Dane from Schleswig-Holstein who assumed duty after Zimmermann, to be 'an excellent man, not only as a ruler but in the way he aids the spread of Christianity … It is a joy to be a missionary to him because he respects his teachers in all ways and protects them.'[34] Although, in time, most of the ruling families became converts, throughout his life Moiloa refused to accept conversion. He was thus able to keep a foot in the camps of both the traditionalists and the Christian converts in his society. Possibly it was because of his reluctance to fully embrace Christianity that the missionaries, even though they wrote glowing reports about Moiloa, were slow to make inroads among the people they sought to convert. It took Jensen a decade to convert 110 of the baHurutshe (the number of conversions did, however, pick up pace in the last decade of the century).

FIGURE 2: Moiloa with missionary Ferdinand Jensen c. 1865 (the only known photo of Moiloa II)

Source: BaHurutshe Tribal Office, Dinokana

It was the missionaries' role in land acquisition in the western Transvaal that was to leave a more permanent and, for most African communities, more important mark. Most historians have remarked on how land purchases in the north-western Transvaal were an important way in which Africans resisted being drawn into wage labour and retained a hold over land when others were losing it.[35] Although the law strictly forbade Africans from purchasing land and holding title to it they could get around this legislation by purchasing land in the name of the missionary, who held it in 'trust' for the chiefdom. Moiloa first acquired land in 1867, with missionary intervention, purchasing two farms next to the baHurutshe location called Dam van Matsego and Matjesvallei. The latter was bought from a well known hunter in the district, Marthinus Swart, for a hundred head of cattle, in apparent contravention of a Volksraad resolution, so it is not clear if this was a legally approved transaction – Moiloa tried in 1874 to get a binding deed of sale from President Burgers but gained nothing before his death. Moiloa also entered into grazing agreements with some of the surrounding Boers, on the farms Welbedacht, Nooitgedacht, Tweefontein and Stinkhoutboom, and he gained access to cattle posts and arable land in Ngwaketse territory. The baHurutshe did not engage in land acquisition on the scale that some other *merafe* did, in particular the baFokeng, but this can be attributed to the fact that Moiloa's location was quite extensive and they enjoyed ensured tenure to it (see Map 7).

Moiloa was an astute politician. On the one hand, he obeyed the Boers and satisfied their demands; on the other hand, in return, he insisted on maintaining a degree of independence. This policy helped to create the kind of stability needed for economic security. The arrival of the trekkers heralded the expansion of a mercantile economy in the western Transvaal hinterland, centred first on hunting and trading in captives and then on land acquisition. By the mid-1850s the extension of the trade frontier in the bushveld region occurred along two routes: one was from Potchefstroom (the commercial capital of the western Transvaal until the last decades of the century) through to Rustenburg; the other along the hunters, missionaries and traders 'road to the north' through Vryburg, Kanye (the baNgwaketse capital from 1852), to Shoshong and then to Ngamiland and Barotseland. A quite popular deviation was to go via Zeerust and up the Ngotwane River to Mochudi in modern Botswana. This meant that a number of mainly English traders such as Chapman, Anderson, Baldwin and Cumming passed through the Hurutshe capital at Dinokana.

In addition, many Boers, acting as middlemen for various agencies in the Transvaal, came to Dinokana to purchase hides, livestock or feathers. Zeerust in 1867 was described as 'a new village in the vicinity of friendly [native] tribes who live in peace and carry on an extensive trade in ivory, ostrich feathers, etc.'[36] The supply of 'exportable produce' from the Marico and Potchefstroom districts was so great that it drew traders from the Cape and Orange Free State, leading the *Transvaal Argus* to complain about the presence of 'colonial sharks hovering about our borders'.[37] The long-held view that the Boers were an isolated community clinging to a subsistence economy has in recent times been exposed as a fallacy[38] – as the activities of the Marico Boers disclose, they were well aware of the commercial potential of southern Africa's interior and were linked to long-distance markets. However, in the western Transvaal the once lucrative ivory and ostrich-feather trades declined as the century wore on, and the Boers were unable to command good prices for their products at the colonial markets of the Transvaal and the Cape. The Boers did, however, stimulate trade and commerce among the western Tswana groups such as the baHurutshe.

The extent and diversity of baHurutshe economic activity between 1860 and 1880 is fully recorded by the missionaries and other visitors to their capital. On his arrival there in 1859, for example, Zimmermann described:

> … this wonderful wide valley with Linokana just about in the centre. The land brings enormous amounts of corn as the people concentrate more on agriculture. Linokana is surrounded by many large vegetable gardens. The Bahurutsi already know how to irrigate. They are generally well off, some even really wealthy, because they have their cattle farming as well as good lands. They have bought many wagons and ploughs.[39]

FIGURE 3: The Dinokana valley with Gopane village in the distance

Source: The authors

A few years later, W Behrens, the HMS supervisor for Bechuanaland missions, reported approvingly of conditions in the town, noting that:

> The baHurutshe have so much corn as they have not had for years. Here in the town of Moiloa there are about five wagons and 200 oxen. You can easily imagine how much work can be done with them. In addition, chiefs, deputies and all who own oxen use the plough, and sow wheat, like the Boers. They hunt in great numbers and shoot wild meat and ostriches and bring back on their pack oxen, meat skins and feathers … for their own use as well as for sale. Here in Moiloa's stad are several thousand guns; a man without a gun is a poor man.[40]

Hunting declined in the mid-1870s owing to competition from white hunters, a decline in the numbers of animals locally, and a restriction by the authorities limiting the baHurutshe to hunting within their reserve. But there was no such reduction in agricultural activity if the reports of literate observers are to be believed. In 1875 the Austro-Hungarian botanist, Emile Holub, recorded that the baHurutshe 'gathered in as much as 800 sacks of wheat, each containing 200 pounds and every year a wider area of land is brought under cultivation … beside wheat they grow maize, sorghum, melons and tobacco.'[41] He also noted that irrigation had become a widespread practice.

The use of wagons indicates that much of this produce was transported elsewhere for sale, and that wealthier producers were investing their profits from agriculture to increase their share in the trading economy. Wagons assumed great significance in the period before white and Indian traders settled in the reserve and bought up locally produced goods. They became all the more crucial from 1869 when a significant new market was opened up at Kimberley by the discovery of diamonds. Holub wrote in 1872 that the baHurutshe 'sold what they did not require for their own consumption in the markets of the Transvaal and the diamond fields'.[42] The diamond fields opened up opportunities for Africans in the western Transvaal bushveld, and from even further afield, to seek the higher wages offered there.

This economic growth benefited many Hurutshe cultivators who were transformed into a thriving peasantry, much as happened among nearly all African societies that engaged with the colonial economy. It also allowed for personal accumulation by Moiloa himself – it is absolutely clear that from his position as *kgosi* he entrenched his own wealth and power through control over land, production and trade among the baHurutshe. He ensured that missionaries and traders purchased grain directly from him (his death, according to the local *landdrost*, had a 'negative effect on trade with the white population').[43] In addition, he was paid £25 per annum by the state to collect taxes from his followers, a task normally entrusted to the *landdrost*, but perhaps a form of reward to Moiloa for his considered loyalty. Possibly the most visible sign of Moiloa's power was the fact that he married eleven women; in the opinion of Holub, who travelled extensively through the bushveld, this was more than most Batswana *dikgosi* at the time. He also formed six new Hurutshe *mephato* (age-regiments) during his period of rule, the largest, Matshelaphala, under his direct control – an indication of chiefly control and political stability.

It should be borne in mind that Moiloa was a 'pretender' to the chieftainship of the baHurutshe. We raised this point in the Introduction, when discussing Moiloa's reluctance to return immediately to his former homeland in the later 1830s, but the issue is complicated and requires repeating; and in addition some of the circumstances prevailing in the earlier period had changed. Sebogodi, the rightful *kgosi*, had been killed in action against the baNgwaketse. He had three sons: Menwe, Motlaadile and Moiloa. But Sebogodi's rightful successor, Menwe, had predeceased his father so Sebogodi's brother Diutlwileng assumed the chieftainship. He, in turn, died during the disturbances of the *difaqane*. This is when Sebogodi's youngest brother Mokgatlhe took over the reins of power. Strangely, Motlaadile seems never to have made a bid for chieftainship and was eclipsed by the Mokgatlhe/Moiloa faction (this may be because when the majority fled Kaditshwene in 1821-22 Motlaadile had remained and became a tributary of the amaNdebele, thus losing support). According to Hurutshe genealogies, Motlaadile had no children. However, to complicate matters Mokgatlhe had married the appointed 'great wife' of Menwe and according to custom had 'raised up seed' of behalf of his nephew. Mokgatlhe's sons Lentswe and Gopane were thus considered by many to be the rightful line of succession once Moiloa died. Others, however, claimed that Moiloa's son, Ikalafeng, should become the *kgosi*.

Even though the laws of succession were neither fixed nor binding among the baTswana, Moiloa could not ignore the fact that he lacked a really legitimate claim to leadership of the *merafe*, and it was a source of obvious concern for him. He countered it by gathering around him diverse groups of supporters on whom he counted to balance the scales of power in his chiefdom. First among these were the Griqua converts whose support he had sought and acquired during the late 1830s and early 1840s. As Moiloa probably anticipated, they offered a ready link with the missionaries and other African converts, and provided a number of services through their agricultural and linguistic (Dutch-speaking) skills. Moiloa allocated them two separate wards in Dinokana where they enjoyed a measure of autonomy and gave him their support.[44] Interestingly, a number of so-called 'coloured' families, presumably descendants of these original Griqua, assumed his name, which they still bear today. He also drew support from converts, and from a number of non-baHurutshe immigrants, who came to settle in the reserve during the 1850s. When Moiloa died, the authorities of the incoming British administration considered that the 'pure' baHurutshe were the followers of Gopane, and that it was the immigrants who enjoyed status at Dinokana. This development indicates also that the ethnic composition of the baHurutshe was constantly changing and was never fixed or immutable – certainly the ethnic face of the pre-*difaqane* and post-*difaqane* Hurutshe *merafe* was quite different, though there was probably a significant core of 'pure' baHurutshe among it. The region had been occupied by different groups of people in the years of Hurutshe displacement, some of whom did not trek away with Mzilikazi, and would have most likely been incorporated into baHurutshe society as the Tswana ward system allowed for the incorporation of 'foreigners' or strangers, who were semi-independent. However, the prevailing trend among minority factions from the mid to late nineteenth century (and perhaps beyond) was to seek for incorporation into the expanding Hurutshe chieftaincy rather than establishing an independent identity.

Though dependent on a number of different sections of baHurutshe society, Moiloa tried at the same time to weld his community together through a process of political involution, (around the person of himself as the *kgosi*). It was therefore essential to maintain and even strengthen key social practices and institutions. His refusal to convert perhaps indicates his desire not to alienate traditional elements of baHurutshe society, as does the re-formation of former *mephato* and the introduction of new ones. In addition, during his time no marriages (save Christian ones) were considered legal without the passing of *bogadi* cattle (bridewealth payment).[45] Moiloa's 'domestic policy' therefore was a dual (though perhaps contradictory) one of strengthening support among 'non-traditional' elements among the *morafe* while at the same time rebuilding the essential props of traditional society that had broken down during the *difaqane*.

Moiloa's external policies were geared to ensure stability with neighbours and with the authorities of the South African Republic. He had a difficult relationship with Setshele's baKwena across the border, who initially regarded Moiloa as a vassal and tool of the trekkers, but he sent Setshele oxen, probably as a token of friendship, and many Boers thought that Moiloa 'would stand by' the baKwena during the war scare time of 1856,[46] an allegation that Viljoen was forced to investigate but later repudiated. His relations with the baNgwaketse also appeared good – he had, after all, earlier sought and received sanctuary from them. According to evidence given in 1871 to the Bloemhof Commission, which sat to arbitrate the various contested claims to the diamond fields, the cattle belonging to the respective *merafe* used to 'depasture' in each other's territory between the winter and summer months.[47] The only nearby group with whom Moiloa did not mend relations was the baTlhaping. When the baHurutshe returned from

Modimong after seeking sanctuary there during the *difaqane*, their *kgosi*, Mahura, was incensed that they had 'defected' to another authority by returning to what became the Transvaal. Mahura consequently dispatched a raiding party into the Transvaal which overwhelmed the unsuspecting baHurutshe killing over fifty people, and returned with a number of cattle.[48] After that Moiloa remained suspicious of the baTlhaping.

Moiloa was also considered by the South African Republic authorities to be a major figure in the politics of the wider region. In 1870 he attended a joint meeting of the baRolong, baNgwaketse, baHurutshe and Kora leaders with, among others, President MW Pretorius, to affirm the territorial integrity of the baTswana bordering the Republic, and he was also invited to give evidence to the Bloemhof Commission.

In July 1875 Moiloa's death was imminent. Although he had steadfastly refused to accept conversion, even on his deathbed, in a typical gesture of compromise he instructed Jensen to lay him in a coffin upon his death and not to bury him in the seated position as was African custom. This symbolic rejection of tradition 'stilled the large group of mourners into silence'.[49] After his death the relative stability the baHurutshe under his rule had enjoyed for nearly two decades collapsed, and a period of division, civil strife, and ultimately dispossession ensued. Moiloa, however, had done much towards creating the necessary unity required for the re-building of the baHurutshe after the turbulent years of the 1830s and 1840s, a role which earned him the accolade of the 'mighty man with thick neck who does not walk behind the people', in Hurutshe praise poems.[50] He was succeeded by Ikalafeng Moiloa in 1877.

OTHER LESSER-KNOWN BATSWANA LEADERS OF THE NINETEENTH CENTURY

Ratheo Monnakgotla

Whereas Moiloa was perhaps the most successful of the lesser-known nineteenth-century southern African leaders, there were others who re-established control over their communities and laid the basis for economic and political security. One was Ratheo Monnakgotla of the baKubung. Towards the end of the *difaqane* he seized control of the fragmented chiefdom from Lesele Mathope, who probably had a greater claim to it. Monnakgotla led his followers to Heilbron in the Free State where they remained for close to forty years. In 1880, the farm where Lesele Mathope had settled at Molotestad, north-west of Ventersdorp, was put up for sale. The two factions seem to have buried the hatchet by this time and Lesele approached Monnakgotla with a view to re-uniting the baKubung and settling as one community at Molotestad; it was to Monnakgotla, however, who actually clinched the purchase of the farm with the assistance of the Anglican missionary to the Bakubung-ba-Mathope at Molotostad, the Reverend Clulee of the Society for the Propagation of the Gospel. Clulee was led to believe that Monnakgotla was the legitimate chief, and assisted him in negotiations for the purchase. Monnakgotla then returned to the western Highveld in 1881. The aggrieved Lesele, who by 1882 had fallen foul of the colonial authorities, abandoned Molotestad with a handful of supporters but some years later they, too, acquired land in the Derby district and called the settlement Mathopestad. The two baKubung groups remained independent from one another, though both were in possession of good farmland.[1]

NOTES

1   These events are recorded by H.L. Dugmore, 'Land and the Struggle for *Sekama*: The Transformation of a Rural Community, the Bakubung of the Western Transvaal', (B.A. Hons dissertation, University of the Witwatersrand), 1985, and in Dugmore, 'The Rise to Power of the Mmonakgotla Family of the Bakubung,' *Africa Perspective*, 1985, pp. 101-116. See also P-L. Breutz, *Tribes of the Ventersdorp District*, Government Ethnological Publications no. 5, (Pretoria, 1957).

KGAMANYANE

The baKgatla ba Kgafela did not fare as well during these years. Pilane, the *kgosi* during the period of upheavals, fled north to the baLaka of *kgosi* Mapela in the country of the baPedi. He returned in 1837 after the amaNdebele had gone, but he appears to have made a serious enemy of them, and the amaNdebele allegedly raided him again from their new abode in Matebeleland as late as 1842,[1] and according to baKgatla traditions some of his sons were taken captive. Pilane's son Kgamanyane succeeded him in 1850 or 1851. By then the baKgatla resided as tenants at Moruleng (on the farm Saulspoort owned by Paul Kruger); in 1864, Henri Gonin, a Swiss national and member of the *Buitelandse Sending* (foreign mission) of the NG Kerk, was allowed by Kruger to work among them and through Gonin's efforts the baKgatla purchased Saulspoort in 1898.[2] It may have looked promising for Kgamanyane's baKgatla, but the demand for forced labour on Boer farms in the Pilanesberg had become unbearable by 1865, and some baKgatla left the district altogether.

In addition, Kgamanyane's relations with Kruger worsened. Early in 1870, upon Kruger's instruction to his representative official in Saulspoort, HP Malan, a Kgatla work team was inspanned to wagons and carts containing stone boulders and forced to pull them to a dam construction site on an irrigation project of Kruger's within Saulspoort. This caused extreme discontent and, on receiving complaints from the men, Kgamanyane agreed that they should stop working. For this 'misdemeanour,'

FIGURE 4: Kgamanyane
Source: Mphebatho Museum, Moruleng

FIGURE 5: Paul Kruger, Commandant 1865
*Source: National Archives of
South Africa, Pretoria*

Kgamanyane was publicly flogged by Kruger at a meeting convened at Saulspoort in April of that year. In addition to continual Boer demands for Kgatla labour and the extortion of their goods and money over land,[3] the flogging was the proverbial 'last straw'. In a state of understandable anger and distress, Kgamanyane, with at least half of his people, emigrated to Mochudi in baKwena country in present-day Botswana. This was a huge blow for the baKgatla who stayed behind in the Pilanesberg, and they remained in a state of relative insecurity until the South African War, when they managed to recoup some of their losses.[4] We follow the fortunes of the Bakgatla-ba-Kgafela during and after the war in the next chapter.

NOTES

1   Breutz, *Tribes of Rustenburg*, p. 257.
2   For an account and analysis of Gonin's life and work among the baKgatla see B. Mbenga, 'The baKgatla baga Kgafela in Pilanesberg District of the Western Transvaal, from 1899 to 1931', Ph.D thesis, University of South Africa, 1996, Ch2; and B. Mbenga and F. Morton, 'The Missionary as Broker: the Rev. Henry Gonin, the BaKgatla of Rustenburg District, and the South African Republic, 1862-1922', *South African Historical Journal*, 36, (1997), pp. 145-167.
3   Morton, *When Rustling Became an Art*, p. 101.
4   For a full account of this incident see B. Mbenga, 'Forced Labour in the Pilanesberg: The Flogging of Chief Kgamanyane by Commandant Paul Kruger, Saulspoort, April 1870', *Journal of Southern African Studies*, vol. 23, no. 1, March 1997, pp. 127-140.

THE REVEREND HENRI GONIN, MISSIONARY TO THE BAKGATLA BA KGAFELA

Henri Gonin was born in Switzerland in the early 1830s. In 1860, he received his theological training in Geneva, Switzerland, and in Edinburgh, Scotland. Before the end of his training in Edinburgh, he was recruited as a missionary by the Reverend Dr DW Robertson of the Dutch Reformed Church (DRC) in South Africa. The foreign sub-committee of the DRC, which catered for black people in the Transvaal and beyond, sent Gonin to open a new mission in the Transvaal. He and his wife arrived in Rustenburg by ox-wagon in May 1862. After two years of waiting in Rustenburg, they went to work among the baKgatla ba Kgafela in the Pilanesberg, with the permission of the most senior government official of the district, Commandant Paul Kruger. The Kgatla chief, Kgamanyane, allowed the Gonins (now with three children) to settle at Saulspoort, his headquarters. In June 1864, in order to have personal independence and also security of tenure, the Gonins bought their own farm, Welgeval (or Welgeval-len), close to Saulspoort. Gonin was on good terms with Kgamanyane and quickly learnt Setswana but struggled to make converts among the reluctant baKgatla. There were, however, a large number of literate Africans living at Saulspoort – the *oorlams*. These were the men and women who Gonin recruited as his teacher-evangelists and depended upon for opening up new mission stations and spreading the gospel all over the Pilanesberg and up to the border with Bechuanaland. As the baKgatla were a people divided by an international border, their paramount chief living in Mochudi, Bechuanaland, Gonin opened another mission station there at the beginning of the twentieth century and converted many baKgatla to Christianity. Apart from his missionary work, Gonin also assisted the baKgatla to purchase land by registering it in his name, as during the nineteenth century Africans in the Transvaal were not allowed to register land in their own names. In 1910, Gonin had become the longest-serving DRC minister in South Africa, having served at Saulspoort continuously for forty-six years.

FIGURE 6: Henri Gonin's original church at Saulspoort, being renovated, 2014

Source: The authors

ENDNOTES

1   Shillington, *Luka Jantjie*.
2   Molema, *Montshiwa*.
3   B. Mbenga and A. Manson, '*People of the Dew': A History of the Bafokeng of the Rustenburg District, South Africa, from Early Times to 2000* (Johannesburg: Jacana, 2010), 27-74; J. Bergh, 'We Must Never Forget Where we Come From': The Bafokeng and Their Land in the Nineteenth Century (*History in Africa*, 32 (2005); and G. J. Capps, 'Tribal Landed Property: The Political Economy of the BaFokeng Chieftaincy, South Africa, 1837-1994', D. Phil thesis, London School of Economics and Political Science, 2010.

4   See M. Legassick, 'The Sotho-Tswana peoples before 1800,' in L. Thompson (ed.), *African Societies in Southern Africa* (London: Heinemann, 1969); Manson, 'The Hurutshe in the Marico', pp. 36-43.

5   A. Manson, 'Confict in the Western Highveld/Southern Kalahari, c. 1750 -1820'; and N. Parsons, 'Prelude to the Difaqane in the Interior of Southern Africa, c.1600-1822, in C. Hamilton (ed.) *The Mfecane Aftermath, Reconstructive Debates in Southern African History*, (Johannesburg and Pietermaritzburg: University of Witwatersrand Press and University of Natal Press, 1995).

6   S. Kay, *Travels and Researches in Caffraria* (London 1833), pp. 225-227.

7   J. Campbell, *Travels in South Africa Undertaken at the Request of the Missionary Society Narrative of a Second Journey*, (London: Westley, 1822) vol. 1, p.261.

8   Records of the Paris Evangelical Missionary Society, *Journal des Missions Evangelique*, vol. 8, 1833, pp. 202-203.

9   See Manson, 'The Hurutshe in the Marico region', pp. 66-74.

10 Official correspondence of the London Missionary Society (henceforth LMS Correspondence), Box 23, Inglis to directors, 26 September 1848.

11 This account of the baHurutshe during the Difaqane is taken from Manson, 'The Hurutshe in the Marico District,' pp. 62-90. More information on the relations between the baHurutshe and the Transvaal is available in A. Manson 'The Hurutshe and the formation of the Transvaal state, 1835–1875', *International Journal of African Historical Studies*, 25, 1 (1992).

12 LMS Correspondence, Box 24, Edwards to Tidman, 19 June 1849.

13 LMS Correspondence, Edwards to directors, 9 September 1849.

14 See Manson, 'The Hurutshe', pp. 96-98.

15 LMS Correspondence, Inglis to Tidman, 24 August 1850.

16 Landau uses the baHurutshe's return to the Marico as a case in point for the creation of 'tribes' by Europeans during the colonial period. Potgieter apparently 'stipulated that only "they, the Barurutse'" [sic] could return to Mosega, thus 'blocking add-ons, allies, subordinate lords and so on'. By so doing he allegedly fashioned the baHurutshe into a distinct 'tribal' entity. But the evidence suggests that they were quite heterogenous, nor had Potgieter any coercive means of enforcing such a stipulation. The notion that African communities were simply some sort of helpless '*tabula rasa*' on which Europeans drew tribal maps is one that needs repudiating. See *Popular Politics*, p.121.

17 Morton, 'Slave Raiding', pp. 102-3. 'Captive Labour in the Western Transvaal after the Sand River Convention', in Eldredge and Morton (eds.), *Slavery in South Africa*, (Boulder and Scottsville: Westview Press and University of Natal Press, 1994), p. 175.

18 J. Freeman, *A Tour in South Africa* (London 1851), p. 274.

19 Delius, *The Land Belongs to us: The Pedi Polity, the Boers, and the British in the Nineteenth Century Transvaal* (Johannesburg: Ravan, 1983) p. 35.

20 Cited in Morton, 'Slave Raiding', 107; 'Captive Labor in the Western Transvaal after the Sand River Convention', in Eldredge and Morton (eds), *Slavery in South Africa*, p. 175.

21 For more information on the *oorlam* phenomenon, see the seminal article by P. Delius and S.Trapido, "'*Inboekselings* and *Oorlams*': The Creation and Transformation of a Servile Class,' *Journal of Southern African Studies*, vol.8, no 2, (1982).

22 For an account of these communities see A. Manson and B. Mbenga, 'The Evolution and Destruction of *Oorlam* Communities in the Rustenburg District

of South Africa: The Cases of Welgeval and Bethlehem, c. 1850-1980", *African Historical Review*, vol 41 (2), (2009); B. Tema, *The People of Welgeval*, (Cape Town: Zebra Press, 2005) is an autobiographical account of growing up on the farm.

23 National Archives' Depot, Transvaal Archive (TA) of the State Secretary (henceforth SS), vol.5, r 468/53, Report of Viljoen's Meeting by *Afteraardigden* (Representative), 16 January 1853. Unless indicated all archival sources are contained in the TA section of the SA Depot in Pretoria.

24 For more detail see J.A.I. Agar-Hamilton, *The Native Policy of the Voortrekkers*, (Cape Town: Maskew Miller, 1928), p. 116; and W. Cochrane, *Memoirs of Reverend Walter Inglis* (Toronto, 1887).

25 Cited in R. Lovett, *The History of the London Missionary Society, 1795-1895*, (London, 1899), p. 596.

26 Cited in J. Chapman, *Travels in the Interior of South Africa* (London 1868), p. 88; and State Archives SS vol.5, r517/53 J.W. Viljoen to A.W. Pretorius, 16 April 1853.

27 See J. Grobler, 'Jan Viljoen, the South African Republic and the Bakwena, 1848-1882', *South African Historical Journal*, no 36, (May 1997) pp. 241-249.

28 For example the Swazi and the Ohrigstad trekkers, and in the Zoutpansberg.

29 Grobler, 'Jan Viljoen', p. 247-248.

30 Transvaal Archives, SS vol. 11, r1127/56, J.W. Viljoen to A.W. Pretorius, 30 July 1856.

31 Cited in W. Kistner, *The Anti-Slavery Agitation against the Transvaal Republic, 1852-1862* (Parow 1952), p. 221 from Eerste Volksraad Notule, E.R.V 105, 1858.

32 Cited in Kistner, *Anti-Slavery Agitation*, p. 222, from LMS correspondence, Mackenzie to Tidman, 8 September 1864.

33 Manson, 'The Hurutshe in the Marico District', pp. 142-144.

34 *HMB*, Report from Linokana Station, 1864, p. 189.

35 See G. Relly, 'The Transformation of Rural Relationships in the Western Transvaal', M.A Thesis, University of London, 1978; J. Bergh and H.M. Feinstein, 'Trusteeship and Black Land Ownership in the Nineteenth and Twentieth Centuries', *Kleio*, 36, (2004) pp. 170-193; J. Bergh, '"We must never forget where we come from": The BaFokeng and their land in the nineteenth century Transvaal', *History in Africa*, 32, (2005) pp. 95-115.

36 *The Transvaal Argus*, 7 February 1867.

37 *The Transvaal Argus*, 21 February 1867.

38 See C. Hamilton, B. Mbenga and R. Ross, *Cambridge History of South Africa: From Early Times to 1885*, vol. 1 (Cambridge: Cambridge University Press, 2010). pp. 340-350.

39 Unpublished HMS Correspondence, "Affairs at Linokana Station", F. Zimmermann, c.1859.

40 *Hermannsburg Missions Berichte* (henceforth *HMB)*, unnumbered, 1864, p. 138.

41 Holub, *Seven Years in South Africa*, (Reprint Johannesburg 1975), vol. 1, p. 22.

42 E. Holub, *Seven Years in South Africa*, vol. 2, p. 22.

43 TA SS r 287/77, Report of Landdrost , Zeerust, re. Conditions in Linokana, 13 November 1877.

44 HMB, no. 7, Report by L. Harms, 'Affairs of Linokana Station', F. Zimmermann, p. 2.

45 Manson, 'The Hurutshe in the Marico district', pp. 156-157.

46 TA SS vol. 11, r 1113/56, notule. 9 June 1856.

47 *Bloemhof Blue Book* CA 21/1, Evidence of Gaseitsiwe, pp. 188-189, and M[a]ranyane, 329.

48 TA SS vol.3 r 290/51 Inglis to M.W. Pretorius, 3 June 1851. See also J.A.I Agar-Hamilton, *The Native Policy of the Voortrekkers* (Cape Town, Maskew Miller, 1928), p. 79.

49 *HMB*, no.4, 1875, p. 221.

50 See Breutz, *Tribes of Marico*, p. 31 from songs collected by missionaries of the Hermannsburg Mission, Dinokana, 1906.

CHAPTER 2

The South African War and its aftermath 1899-1908

INTRODUCTION

The South African War is one of the hinges upon which modern South Africa turns. On the highveld, especially, it swept aside the old Boer states and ruling class; it transformed the human geography of the region; and it thrust entire communities into new political and economic relationships. After the war's end, the British set about modernising the South African state.

In the region of the north-western bushveld, the South African War involved every African society in varying degrees, some total and committed participants, others onlookers inevitably and reluctantly swept up in the dramatic episodes of the conflict. In this chapter we explore and analyse the varying roles of the black participants in the South African War and its impact upon them. The particular focus is on the the baKgatla, the baRolong, and the baHurutshe, with passing reference also to minor players such as the baFokeng, baTlokwa and others.

Most of the contemporary or early accounts of the mis-termed 'Anglo-Boer' War perpetuated the myth that it was a 'white man's war' that did not involve Africans. Only comparatively recently was this laid to rest when several historians, in particular Peter Warwick and Bill Nasson, revealed the very active participation by black people in a wide range of roles in the war (including armed combat) on both the British and Boer sides.[1] All these historians have shown that blacks were both 'active shaping agents as well as victims' in the war.[2]

The reasons for the conflict, discussed and debated in many books and academic articles,[3] apply to the major combatants at national level; in the various regions where Africans became involved in the war they did so because of their own local, specific reasons, some of them deep-seated historical grievances.

BaKgatla participation in the War[4]

The baKgatla harboured specific historical grudges against the Boers. One was the Boers' incessant demands for their labour over several decades which finally culminated in the flogging of Kgamanyane by Kruger discussed in Chapter 1. In both the Pilanesberg and Mochudi, this episode was remembered with bitterness by practically all of the older baKgatla men and women who still retain graphic accounts of the flogging and the subsequent division of the *morafe*. It was because of this that they saw the war as an opportunity to take revenge on the Boers, to regain the land in the Pilanesberg which they had lost to the Voortrekkers, and to reunite their divided people.[5]

The decision to involve the Pilanesberg baKgatla in the war was made in Mochudi, the seat and source of their chiefly authority from 1870, by their paramount chief Linchwe. Following the Derdepoort episode (discussed below), the baKgatla anticipated and prepared themselves for possible Boer reprisals[6] as, owing to the strong unity between the baKgatla on both sides of the border, any large-scale conflict facing one section involved the other as well. Every regiment in Mochudi had its counterpart (with the same name) in Saulspoort, the Mochudi being the senior – there were, for example, two Makoba regiments, the Mochudi under Ramono, and the Saulspoort under Ditlhake.[7]

On the eve of the war, the British and the Boers feared that discontented African groups might use the war to advance their own agendas and possibly even attempt to overthrow white rule – this explains the elaborate preparations by both sides to counter any possible African insurrection. The fear was not totally misplaced because many African chiefdoms, including the baKgatla, took the British side, convinced that it would be victorious and that such support would gain them considerable political and economic benefits after the war.[8]

Some five months before the war, official British policy was still against using blacks in armed combat against the Boers. There were two reasons. First, after their defeat in the Anglo-Boer War of 1880 the British were uncertain about black loyalty in the Transvaal. Second, the British considered it morally unacceptable to employ blacks in combat in a war between two white societies. The practical realities of the war, however, were such that the British would need African assistance, but the British authorities were worried about the loyalty of Linchwe, for half of his followers lived under Boer rule and if he took sides it could mean exposing his people to obvious danger. The contemporary writer, LS Amery, expressed Linchwe's predicament aptly when he wrote that 'his [Linchwe's] interests, even in peace time … pulled him two ways'.[9] As the possibility of war became increasingly real, British concern grew with the realisation that the baKgatla were potentially their most dependable military allies, whom they had moreover rated as 'good fighters'.[10] The British needed Kgatla military assistance. Moreover, the railway line which was so essential for the transportation of the (British) Rhodesian troops to the south passed through the centre of Kgatla territory. Consequently, the British decided on a psychological show of force to influence Linchwe into supporting them. Just before the outbreak of war, some 500 British South Africa Police (BSAP) and Southern Rhodesian volunteers were sent to Mochudi.

Initially, Linchwe gave a deliberately false impression of either neutrality or uncertainty, while buying time. This worried the British authorities and prompted Colonel RS Baden-Powell in Mafikeng to report that Linchwe was 'wavering'. While biding his time, and before openly declaring which side he supported, Linchwe gave the Boers in the Derdepoort *laager* the false impression that he was on their side by, for example, sending them gifts of slaughter stock. (Derdepoort was situated on the Transvaal side of the border, on a farm owned by PJ Hans Riekert, commandant of the mounted police, North-Western border. This was also his work station.)

At this stage, both the British and the Boers still envisaged that Kgatla military assistance would be limited to noncombatant roles. Yet both sides nevertheless considered their assistance to be crucial. The baKgatla lived under both Boer and British administrations and both were familiar with baKgatla military valour – the Boers much more so than the British, and Boer expectation of Kgatla military assistance was based on a greater familiarity and much longer mutual historical experience. With the outbreak of Boer-British hostilities, the 'wavering' baKgatla had to make a firm decision about who to support.

Linchwe's decision to join the war was influenced by Boer acts of aggression in or close to baKgatla territory in the Protectorate. These occurred after the siege of Mafikeng, when Boer commandos moved into the Bechuanaland Protectorate and began destroying the railway line linking troops stationed in Rhodesia with those to the south. In late 1899 the towns of Lobatse and Gaborone were occupied by the Boers, and cattle looted from Linchwe's people.[11] At the same time, this seizure of Kgatla cattle also occurred in the Pilanesberg. Such incidents were reported to Mochudi, with the message: 'Linchwe must prepare for war because the Boers are coming to do the same thing there.'[12] Despite all signs of Boer hostility, however, Linchwe was still very cautious and would not show support for either side. Meanwhile, the Boer acts of aggression towards the baKgatla continued.

The incident that finally decided Linchwe and his people to support the British cause was the 'insult' which, as Kgatla traditions widely assert, was publicly hurled at Linchwe by the Boer commandant, PJ Hans Riekert. Riekert was riding through Mochudi at the head of a commando *en route* to cut the railway line when the *kgosi* asked them not to disturb the women and children of the town. Riekert retorted scornfully and declared: 'Your chieftainship is no more than a piece of dust.'[13] While this incident helped to swing Linchwe towards the British he was militarily too weak to retaliate against the Boers at this stage. Therefore, until the British troop reinforcements arrived from Rhodesia, the baKgatla, like the British in the Protectorate, continued to be vulnerable to Boer attacks.[14]

The arrival of Rhodesian troops at Mahalapye in early November 1899, and the news that Chief Khama of the baNgwato had just repulsed a Boer commando attack at Selika Kop gave Linchwe confidence to enter into the conflict. He requested WH Surmon, the assistant commissioner for the southern district based at Gaborone, for arms with which to fight the Boers. In response, at Mochudi railway station the commander of the British troops, Lieutenant-Colonel GL Holdsworth, arranged for the Kgatla regiments to be part of, and led by, the British forces. In mid-November 1899, the baKgatla began to prepare for war.

Three Kgatla regiments, each under Segale (the overall leader of the Kgatla troops), Ramono and Modise, joined the British troops for an immediate assault on the Boer laager at nearby Derdepoort. Official British policy was that the baKgatla were to remain on their side of the border and not to fire unless fired upon first. Moreover, the baKgatla were to be used by the British forces only as guides and carriers.

However, the limitation of the Kgatla role to 'guides and transport assistants' was omitted from the telegraphic instructions sent to Holdsworth on 22 November 1899 by his superior, Colonel JS Nicholson, commandant-general of police, at Bulawayo, Southern Rhodesia. The significance of this omission is that Holdsworth could 'make his own arrangements'[15] as to whether he would commit the Kgatla troops in armed combat, a loophole that Holdsworth was later to exploit with serious consequences for both the Boers and the baKgatla. Moreover, the condition that the baKgatla 'were not to fire unless ordered' in itself presupposed that they could be used in a combative role if necessary – even though official British policy insisted on the exclusion of blacks from armed combat, it nevertheless left a leeway, deliberate or not, for a Kgatla armed role in the impending attack on Derdepoort.

The British-Kgatla attack on Derdepoort caught its white residents unawares, with disastrous results for the defenders. The Boer miscalculation was due to the fact that most men were either stationed elsewhere, were preparing for a

British counter-attack on Mafikeng, or for the capture of Kimberley. It seems that the defence of Derdepoort in the event of a British attack was not a priority for the Boer authorities, General Cronje having misled President Kruger by telling him that Derdepoort was a safe place and suggesting that Boer military attention should be directed elsewhere.

The Derdepoort attack, 25 November 1899, and the Sidney Engers incident

The combined British-Kgatla force under the command of Holdsworth and numbering about 120 men left Mochudi on foot for Derdepoort in the evening of 24 November 1899. The assault on Derdepoort began in the early hours of 25 November. Holdsworth ordered that once the force was near the laager, the baKgatla (and not the British troops) should climb up because, as Ellenberger recorded, 'he [Holdsworth] feared that our men's heavy ammunition boots would betray us when climbing up to the laager and he decided that the bare-footed natives should do the climbing, Segale guiding us to a place from which we could see the laager and open fire on it …'[16] This decision meant that the baKgatla were now very likely to have to fight, contrary to the official British policy. It also clearly shows that Holdsworth was, indeed, using the discretion allowed him by Nicholson.

One Kgatla regiment under Ramono was ordered to climb and secure the ground leading up to the Boer laager across the Marico River, thus effectively crossing into South African Republic territory contrary to strict official instructions. The rest of the baKgatla were on the Protectorate side, while Holdsworth and his men were just inside the Transvaal, strategically placed with a Maxim gun on a ridge on the western bank of the Marico, overlooking the Boer laager. All the troops then began firing. But there were two logistical problems. First, both the British and Kgatla troops were so dangerously placed that when either of them fired at the laager there was the danger that stray bullets might hit their own forces. Second, the British troops could not quickly cross the river because there was no accessible drift. Holdsworth decided to withdraw his troops and return to Mochudi.[17] This provided the opportunity the baKgatla wanted to begin settling their long-standing grievances against the Boers. In the ensuing fight, the baKgatla suffered fourteen dead and sixteen wounded, while twenty Boers were killed, including JH Barnard, a member of the first *Volksraad* for Rustenburg. The baKgatla also captured 100 oxen, thirty horses and eighteen women and children.[18]

From now on, and on their own initiative, the baKgatla began to extend their sphere of activities far beyond Derdepoort. This led to the unfortunate murder of a German trader, one S Engers, in Sikwane village close to Derdepoort. Shortly after the attack on Derdepoort had began, a small group of Kgatla troops under a man named Mongale was instructed by Ramono to take Engers prisoner, believing he was in league with the Boers. Ramono's instruction also stipulated that Engers 'was to have been made a prisoner and not killed',[19] which clearly reveals the baKgatla's intentions. The Kgatla troops, however, did not take Engers prisoner but killed him instead. From an investigation conducted by the British some eight months later it emerged that the Kgatla decision to kill Engers was made on the spur of the moment; when a small group of Kgatla fighters surrounded Engers's shop in the early hours of 25 November and asked him to come out because 'the chief wanted to see him' he was suspicious and incredulous. Upon realising that his shop was surrounded by a party of armed blacks, Engers, fearing for his life, panicked, and allegedly attempted to summon assistance from a number of Boers stationed across the Marico River.[20] According to the baKgatla, Engers 'made for the door ... [and] thinking he was armed, as we were, he was shot down'.[21] According to another eye-witness, Engers's housekeeper, he had been stabbed twice in the shoulder before he was shot. The baKgatla, now numbering some fifty men, looted the shop, destroyed Engers's papers, and stole about £700.[22] Looting and stealing may even have been one of their original intentions.

Unsurprisingly, the murder elicited outrage from the Boers and some of their supporters. The British, however, refused to investigate it, presumably because they regarded the baKgatla as allies and had given them free rein to fight in this vicinity of the country in order to free up British troops to fight the Boers elsewhere.[23]

The attack on Engers, and on other Boer homesteads by the baKgatla, were widely circulated and sensationalised by the Boer community in the bushveld region. For example, it was recorded that Engers 'was disembowelled and otherwise tortured ...'[24] while another report stated that some of the women and children taken captive by the baKgatla were murdered.[25] These impressions of alleged Kgatla brutality were to linger for many years.

The accusations, however, were generally exaggerated and contradicted by a number of sources: the assistant commissioner at Gaborone, WH Surmon, reported that 'not one of [the Boer captives] ever mentioned to me that they had in any way been ill-treated by the baKgatla beyond being required to walk from Sekwani to Mochudi'.[26] Even Engers's housekeeper, who was taken as a prisoner of war, refuted allegations of Kgatla brutality, claiming that she saw no outrage or violence of any kind committed on the Dutch women.[27] This, however, is not to exonerate the baKgatla from all blame. They did destroy Boer property in Derdepoort by setting fire to a number of buildings, including Commandant Riekert's house.

The aftermath of the Derdepoort attack

The Boers felt aggrieved and alarmed by what had happened at Derdepoort. On 26 November, one of the survivors in the attack cycled to Rustenburg and alerted the Boer military authorities. The response was one of panic followed by a desire to exact reprisal. A large contingent of troops was dispatched to Derdepoort from Rustenburg and the Crocodile Pools, while many others were mobilised from the districts of the Heks and Kgetleng Rivers, Waterberg, Mafikeng and even Johannesburg. By the beginning of December 1899, the combined number of Boer troops at Derdepoort had been beefed up to well over 400.[28]

Fuelled by fear and panic, the Boers believed that the baKgatla were planning a massive invasion of not only the tiny white administrative post at Saulspoort, but even Rustenburg itself. Commandant JF Kirsten sent an urgent message to the *landdrost* of Rustenburg, Mr JC Brink, requesting him to call out the entire district commando in defence of the town, and the jail was converted into a fort. Farmers in the district formed themselves into *laagers* and many moved into Rustenburg with their whole families and herds of cattle, for protection. The fears generated by these rumours and, of course, the earlier Derdepoort incident itself, galvanised the Boers into action. Inevitably, revenge was exacted, and on 22 December 1899, in a major engagement, the baKgatla succumbed to the Boers' cannons and Maxim guns and sustained 150 losses to the Boers reported four.[29] In order to punish the baKgatla and ensure that they would not pose any more danger, the Boer commandos razed the Kgatla border villages of Mathubudukwane, Malolwane and Sikwane. They even threatened to attack Mochudi itself. Sikwane, in particular was virtually destroyed by the Boers in case (since it was the closest to Derdepoort) it would be used as a base against them. But instead of achieving their intended objectives, the Boer actions made the baKgatla even more determined to fight back. They took the razing of their border villages as a sign that a state of war existed, and began to escalate it throughout the Pilanesberg. In January 1900, following Segale's request to the British authorities, the baKgatla received 100 more Martini-Henry rifles, in addition to those they had obtained for the Derdepoort attack. The British allowed the Protectorate baKgatla to enter the Transvaal with these weapons, a sure sign of tacit British approval for escalating the conflict with the Boers.

On 16 February 1900, Linchwe mobilised two Kgatla regiments, the Makoba and the Mojanko, under Ramono and Motshwane respectively. At Kayaseput, half-way between Derdepoort and the Dwarsberg mountains, the regiments waited in an elaborately planned ambush for a large convoy of Boer troop reinforcements and supply wagons from Rustenburg, bound for Derdepoort. In the ensuing attack, 'many Boers' were killed, and their wagons and supplies captured. The news of this incident was so unsettling that a Boer commando at nearby

Sepitse abandoned their laager when they heard about it.[30] This famous incident in Kgatla history has vividly lived on in popular memory, especially among the older generation, and is graphically depicted in a praise-poem dedicated to the bravery of one of the Kgatla commanders in that incident, Ramono.[31] Soon after the Kayaseput ambush, Commandant P Steenkamp and some of the Rustenburg commando went to Derdepoort and escorted back the remaining Boers. As a direct result of that ambush, Derdepoort was abandoned for the remainder of the war.

The Derdepoort and Kayaseput encounters produced two results. First, by illustrating that the Boers were defeatable, they boosted baKgatla morale with the confidence to pursue the war more vigorously, as further military engagements were to show. Indeed, in this regard the Boer withdrawal from Derdepoort was highly symbolic. Second, the Boer military threat in the western Transvaal-Bechuanaland border area was reduced considerably as a direct result of the events at Derdepoort, because the Boers then concentrated their attacks south of Gaborone towards Mafikeng, enabling the British to re-occupy Gaborone, which the Boers had earlier forced them to abandon in October 1899.[32]

Further Boer-Kgatla military engagements soon followed. At Moreteletse in the Mabeskraal area, the baKgatla captured 300 Boer trek-oxen and two wagons – but in this encounter Tlatsi, Linchwe's *ntona* (confidential assistant) was killed in action.[33] In mid-1900, with a new supply of 250 Martini-Henry rifles from the British authorities, the baKgatla were clearly on the warpath. In July 1901, Linchwe obtained permission from the assistant commissioner at Gaborone to protect his people at Saulspoort who were in danger of attack, but was told 'not to attack outside his own country [Mochudi in Bechuanaland]' although the message further added that 'there is no reason why he should not send some of his men to assist Saulspoort'.[34] Linchwe was twice supplied with an unspecified number of rifles and ammunition by the British military authorities. This assistance shows the extent of Britsh reliance on the baKgatla.

Operating south of the Kgetleng River, as far as Rustenburg, the baKgatla were so militarily effective that, in the words of the sub-native commisioner at Saulspoort, FE Edmeston: 'The military authorities were relieved of all anxiety as to this district, which was held by these [baKgatla] people, as far north as Palla [Pella].'[35] The triumphant baKgatla now had occupation of all land between the Crocodile and Elands rivers, a no-go zone for the Boers.

During the war in the Pilanesberg the baKgatla looted Boer cattle on a massive scale. As the native commissioner in Rustenburg reported at the end of the war, 'ninety-nine per cent of the cattle looted from the Boers' was by the baKgatla, most of it 'at the instigation and with the cognisance of the [British] military authorities'.[36] There were very few instances of other ethnic groups in the Pilanesberg looting Boer cattle – but this was only to be expected as the baKgatla were the Boers' only adversaries in this area during the South African War. BaKgatla looting of Boer cattle was in all likelihood beyond the control of the British authorities. The looted cattle were carted off to the safety of Mochudi because, as one Mokgatla source put it, once they were there 'it was the end of them, whatever the Boers did, they could never get them back'.[37]

The baKgatla did not only loot the Boers' cattle. They considered other Tswana groups in the Pilanesberg fair game – for example, they raided stock worth £828 from the baPhalane.[38] In November 1902, the native commissioner reported that the baKgatla had 'looted stock from most of the natives residing within reach of Saulspoort' – he had received 'numerous complaints from various tribes' in the Pilanesberg district whose cattle had been 'raided or stolen from them by the Bakhatla (Lintsue's people) during the war', and to ensure the cattle's safe-keeping 'ninety-nine per cent of it' was taken away to the Protectorate.[39] The baKgatla presumably rationalised these as punitive actions against people they considered to be collaborators with the Boers. If chief Linchwe's report of 14 February 1902 to the resident commissioner in Mafeking is to be believed, some African groups 'on the Marico and thereabouts are helping the Boers and are supplying them with information'.[40] Moreover, during the war, as the collective Kgatla leadership itself countered, Boer forces gave their cattle to some friendly baTswana groups in the Pilanesberg (the baFokeng of chief Mokgatle, the baKwena of chief Lerotoli, and the baPhalane of Ramokoka) for safekeeping and acted as scouts for them.[41] These allegations provided the baKgatla with a convenient motive for attack.

The extent of the baKgatla destruction of Boer property in the Pilanesberg was in fact much greater than has hitherto been acknowledged in any of the current historiography on the war in the western Transvaal, even in the works of Morton and Krikler.[42] Travelling through the Pilanesberg just after the war, a British journalist, EF Knight, saw numerous cases of the destruction of Boer property, 'not by our [British] troops – for the British columns never operated in this region, though they occasionally skirted it – but by the … baKgatla tribe, whose work of destruction was far more complete than that effected by our own soldiery'.[43]

The Boers counter-reacted. The period from August to December 1901, which saw an escalation of Kgatla activity, was also the time in which Boer retaliatory raids were at their peak. In a major battle on 12 December 1901, for example, a Boer commando under Commandant JCG Kemp attacked Saulspoort itself and continued raiding northwards up to the Bierkraal River. In the process, some thirty herdsmen under one Kgaboesele attacked the commando 'on its retirement' until they (the herdsmen) ran out of ammunition. Kgaboesele and five others were killed, and the rest were wounded. Apart from the dead and wounded, this engagement was very significant because it 'cost the tribe some 6 000 to 7 000 head of cattle, without mentioning small stock ...'[44] Although this baKgatla claim may have been exaggerated it nevertheless points to the severity of the Boer assault and the incident shows that the baKgatla could not take the Boer forces for granted.

At the end of the war, the sub-native commissioner reported that there had been 'summary executions' as well as 'numberless cases of cruelty by Boers [which] took place in public at Saulspoort, amongst which the flogging to death of an invalid, Mogasoe Segogoane, is the worst case'.[45] The Boers also flogged anyone they suspected of giving information to the British, and 'men and women were wantonly shot down when ploughing their lands in the Pilanesberg'.[46] In contrast, as the sub-native commissioner claimed, the baKgatla 'never mutilated the bodies of the enemy, nor injured a Boer female or child'.[47]

By that stage the baKgatla had considerable success in achieving some of their objectives of the war. They had driven practically all the Boers off their Pilanesberg farms, and occupied them. Although this turned out to be only short term, it was nevertheless symbolic of the strong Kgatla desire to re-occupy what they considered to be their ancestral lands. They had also succeeded in looting numerous herds of Boer cattle which, unlike the temporary occupation of Boer farms, became a permanent and useful community resource.

Consequences of the war

The looted Boer cattle more than compensated for the cattle the baKgatla themselves had lost to the Boers. Cattle were the all-important resource that the baKgatla *kgosi*, Linchwe, used not only to boost his status and prestige but also to buy his people in the Pilanesberg much-needed additional land. With the Boer defeat, Africans in the Transvaal generally believed that the Boer farms they had occupied during the war 'would be confiscated and given to the natives', but the British authorities 'lost no time in dispelling this delusion ...'[48] and ensuring a quick return to the *status quo ante bellum*. Since one of the Kgatla war objectives was to regain possession of land they had formerly occupied they were greatly disappointed when, following British victory, they were ordered to vacate the Boer farms. The sub-native commisioner wrote that:

> ... the Natives are greatly disappointed at not being made grants of land in consideration of the services they rendered to our troops during the late war; they fully expected that the farms would be taken from the Boers and given to them. They anticipated the [Boer] farmers being dispossessed of all title to land.[49]

There are no statistics of Kgatla war losses, but they were significant. Morton suggests that the baKgatla probably suffered a higher casualty rate because the Boers were 'more experienced with the rifle and therefore a better shot',[50] and has estimated the number of Kgatla men killed in action at some 200 in addition to the many who were wounded. Then, the insecurity of war conditions and the absence of men in military action would have prevented normal agricultural activities from taking place. In the late 1920s, a survivor put it starkly when he said: 'All that they [baKgatla in the Pilanesberg] knew was famine. It was famine that scattered them hither and thither and not war.'[51]

Relations between the Boers and Africans were transformed, at least in the short term, as a direct result of the war. Krikler, who has interpreted Kgatla behaviour as a form of 'revolution from below', has recounted how many farm workers all over the Transvaal deserted their Boer masters during the war, never to return to 'their exploiters'.[52] Even more striking was the changed nature of the Africans' attitudes to the Boers and to whites in general. Through the war the Africans had become, as the Boers and the British administrators put it, 'disrespectful' to white people.[53] These attitudes were even more pronounced among the baKgatla, who had played a major role in the Boer defeat and, as Krikler has correctly stated, their struggles 'were probably the most effective and militant of all those waged by rural working people during the South African War'.[54] The baKgatla clearly saw their war role as having placed them on a different and special level, and they therefore expected to be treated accordingly – which was regarded in some British quarters as conceit. When the promises made during the war proved false, many baKgatla showed their displeasure by refusing to work on white farms or for the incoming British administration.

One of the immediate objectives of the post-war reconstruction administration of the Transvaal under Lord Milner was to restore the pre-war social relations between black and white. This necessitated revamping the administrative framework in what became the Transvaal colony. A Native Affairs Department had even been formed by the British in 1900, during the war period. At its head was a commissioner for native affairs and below him the secretary for native affairs, a permanent head of the department. On 1 July 1904 most of the bushveld's black population fell under the western division, based at Rustenburg, under whom were the separate districts of Pilanesberg, Marico and Rustenburg, each with a sub-native commissioner in charge.

In addition, a South African constabulary was formed in June 1910. This was a military force of 7 500 men, mostly from Canada, Australia and Britain. Commanded by Major-General RS Baden-Powell, it was a rural police force with a network of police posts throughout the Transvaal (including the Pilanesberg) and the Orange Free State. The constabulary's major purpose was to deal with the threat of African rebellion and insubordination in the two Boer colonies. It was used to round up and coerce black workers to return to their pre-war employers, to protect the Boers and their property and to 'promote their [Boer] material interest in the immediate post-war period'.[55] In the bushveld region generally, the South African constabulary played a crucial role in providing physical and psychological security to the very few courageous Boer farmers who did return at the end of the war.

Another important means of restoring pre-war racial inequalities was to disarm the African population. The baKgatla, like other Africans throughout the western Transvaal, were ordered by the British authorities to surrender their firearms soon after the end of the war, but many secreted their weapons across the border in Bechuanaland, and there does not seem to have been any real intent to disarm them. In 1903, the high commissioner remarked on the 'loyal attitude' of the baKgatla during the war, for which reason he had decided not to force them to return the Boer cattle they had looted,[56] and this was probably the reason they also retained their rifles.

The major preoccupation of sub-native commissioners in Rustenburg and the Pilanesberg immediately after the war was the handling of African claims of compensation for war losses.[57] Linchwe claimed compensation as early as 26 November 1901, well before the formal end of the war, when he wrote to the assistant commissioner in Gaborone, asking for £10 248 13s for 'my people in the Transvaal, whose cattle, sheep, goats and horses were taken by the Boers during the present war'.[58] Linchwe was told, however, that the Transvaal baKgatla would be treated in the same way as other loyal Africans who had also suffered losses. In view of the baKgatla military role and the favourable treatment which Linchwe expected in return, he must have been disappointed by this rebuff. It must also have been difficult to come to terms with the apparently contradictory attitude of the British authorities, who on the one hand regarded them as special allies and on the other forced them to comply with the regulations applicable to all Africans in the Transvaal.

Another example of ambiguity was about the order that Africans should return looted Boer cattle to the rightful owners, from which the baKgatla were seemingly exempt – a very half-hearted effort was made to round up looted cattle from them, and only a hundred head were returned to white owners (it was of course difficult and impractical for the Boers to follow up and identify cattle in the Protectorate in any case). From their side, the Boers refused to give up cattle claimed by the baKgatla.[59] The matter was finally put to rest by the high commissioner's decision that, because the baKgatla's 'loyal attitude' during the war had not been rewarded by the British it would be 'an ungenerous act to take from them the comparatively small share of plunder with which they have recouped themselves for losses sustained at the hands of the Boers'.[60]

For the baKgatla, the war was important in shaping their future. The enormous herds of cattle they had looted more than compensated for their earlier losses during the rinderpest epidemic a few years previously and the forced migration of thousands of followers of Kgamanyane in the 1870s. They fully exploited the opportunities the war presented to re-build their cattle stocks and after the war these cattle became extremely important as a resource with which to buy badly needed additional land in the Pilanesberg. This, in turn, boosted Linchwe's chiefly authority among his people. Politically, Linchwe gained enormous prestige and authority among the baKgatla on both sides of the border and was able to install his brother Ramono as chief in 1902, despite British official resistance. The same looted cattle also contributed to general Kgatla prosperity that lasted for almost two decades after the war. The journalist, EF Knight, who visited the area in 1903, observed the following about the baKgatla heartland, Saulspoort: 'Many of the leading men live in well-built houses of red brick. Signs of considerable prosperity and a relatively civilised condition are everywhere apparent.'[61]

The baKgatla failed to attain the important objective of repossessing their ancestral land in the Pilanesberg. At the end of the war, they still had the same four properties as at the beginning: Saulspoort 38, Modderkuil 39, Kruidfontein 40 and Holfontein 361 (they did, of course, 'possess' the abandoned Boer farms for the short duration of the war). The new British administration ensured that the pre-war political and social relations prevailed once again, and the baKgatla became greatly disillusioned. However, although after the war the baKgatla were forced to vacate the Boer farms, many Boers did not reoccupy them until after the First World War and most Boer farmers generally kept out of the area. For some eleven years from the end of the war there was ample grazing land for Kgatla cattle.

The baHurutshe

There are some similarities between the experiences of the baKgatla during the war and those of the baHurutshe. As described in the previous chapter, the baHurutshe were located in what was called Moiloa's Reserve in the Zeerust district. Their reserve adjoined the border with Botswana, running north from Zeerust for approximately 150 kilometres. We argue that, like the baKgatla to their east, the baHurutshe profited from the war and its immediate aftermath to increase their political cohesion and to expand their rural economy.

The western Transvaal was the scene of many of the war's more dramatic episodes and the baHurutshe witnessed them from close up. Hurutshe involvement in the war took three forms. First, like the baKgatla they also seized the opportunity to arm themselves on a scale unprecedented since the 1840s. In May 1901, a new resident magistrate in Zeerust, Charles Levey, was appointed to replace the former Boer evacuee. He moved to the Hurutshe capital of Dinokana in October to be of greater assistance to the military forces under Lord Methuen which were attempting to exterminate Boer commandos in the Enselsberg region which had on a few occasions raided the villages of Dinokana and nearby Gopane for supplies. Levey organised for the arming of the 'loyal natives on this border, for their protection and to watch for every movement towards the railway line'.[62] The number of guns given out must have been substantial for the Hermannsburg missionary among the baHurutshe, Ferdinand Jensen, wrote (perhaps somewhat exaggeratedly) that by October 1901 'all the blacks in Dinokana are armed'.[63]

The second form of Hurutshe involvement in the war stems from this supportive attitude by the British authorities. Hurutshe activity in the war was very similar to that of the baKgatla in the Pilanesberg, although the latter were much more involved. The baHurutshe plundered openly from Boer homesteads in the Marico due to the *carte blanche* given by the authorities to do so. In late 1900, Chief Gopane was summoned to Mafikeng by a Captain Harding who instructed him to loot freely from Boer farms in the district, a command issued again in 1901 at Manoane by a Lieutenant-Colonel Churchwood, commander in the Marico. Armed and under military orders to loot, Gopane's people took the opportunity to capture cattle, horses and donkeys from the Boers, a total number of over forty-eight animals. Two of Gopane's followers lost their lives in these looting expeditions.[64]

The third aspect of Hurutshe involvement in the war was participation in military skirmishes. When the war broke out the baHurutshe hedged their bets and appeared to support neither side, unlike the baKgatla who, though equivocal at first, were quicker to back the British. However, according to a local storekeeper, one Southwood, the baHurutshe 'leant towards the Boers' in 1900 and were incensed at him for 'leading in' Colonel Plumer's column from Ramatlabama to Mafikeng. It seems that at least some baHurutshe did support the Boers, either by choice or by coercion. Jensen, for example, reported that the baHurutshe 'had the reputation of being allies of the Boers.[65] Some support for the Boers, it is safe to suggest, was also located around the mission station in Dinokana. The missionary families actively supported the Boers: Jensen's son Thomas served as a telegraphist for the Marico Boers besieging Mafikeng, and his son-in-law Christoph Penzhorn (whose father was a missionary in Phokeng) served as an adjutant with General Cronje and was captured at Pardeburg in 1900.[66] Some of the mission converts might well have adopted such pro-Boer sympathies as in Phokeng, where converts sheltered and fed Penzhorn and a Boer commando.

One of General Snyman's last acts before departing for the field was to call up forty Hurutshe men who were to report to the *landdrost* at Lichtenburg to guard or work on the farms of *burghers*. More men were called up over the next six months, mainly to safeguard farms. When the British gained supremacy in the region they summoned Hurutshe chiefs to Mafikeng in the latter half of 1900 and gave instructions that the baHurutshe should prevent Boer commandos from raiding the railway line between Mafikeng and Ramotswa in Bechuanaland – a task which they were reported to have performed admirably.[67] The importance of the baHurutshe to the British war effort in the western Transvaal was emphasised by Levey's relocation to Dinokana in 1901 to 'stiffen the resistance' of the baHurutshe, now becoming increasingly alarmed by the proximity of starving Boer commandos who viewed the good crops in the location as a potential commissariat. In December 1901, about a dozen Boers rode into Dinokana to obtain food. A skirmish ensued during which Pringle was captured, then rescued by a group of armed baHurutshe. Six Boers were wounded, one of whom collapsed dramatically in Jensen's house where he later died. The Boers threatened to retaliate by razing Dinokana but they were too disorganised and enfeebled by this stage to carry out such threats. No wonder then that the baHurutshe's efforts in the war were considered to have been 'exemplary' by the British authorities who gave close consideration to the various requests for compensation made by Hurutshe chiefs after the war.[68]

The circumstances of war gave several individuals the chance to settle a few old scores and to exploit some favourable opportunities. For example, Marruran Molotsane, a moHurutshe resident on a farm of the Sephton brothers near Zeerust, had eighteen cattle looted by followers of Chief Shuping's baHurutshe just across the border. On investigation, the authorities discovered that Marruran had been entrusted with four cows but had not returned them. Such acts of theft between Africans, reflected also in baKgatla seizures of African-owned cattle in the Pilanesberg, impinge to some extent on Krikler's view of the South African War as a time when dispossessed blacks united in an attempt to topple their former masters, the Boer landowners. Moreover, the Hurutshe leaders acknowledged the neutrality of 'loyal' Boers in the Marico, and left them alone or promptly returned looted cattle to them once their status had been made clear by the British authorities.[69] Immediately after the war, in October 1902, the native commissioner for Zeerust reported that fifteen Boer families 'returned to the Dionokana Stadt and have taken up their abode there having prior to the outbreak of war been granted occupation there'.[70] Their safe return suggests that a fairly close relationship still existed between certain Boer families and their black neighbours. But, not surprisingly, Ernest Stubbs, the native affairs commissioner, regarded the granting of refuge to Boer families by African societies as a 'dangerous precedent'.

There were instances in the Marico of vengeful attitudes. On the farm of the Hamms (a German family of missionary origins) at Abjarterskop, just after the commencement of the war, JHE Hamms was shot dead by two of his Hurutshe farm servants who, in the company of another farmhand, looted the farmhouse.

The evidence given by witnesses at the trial indicates that the act was linked to the general disorderliness of the time created by large-scale looting by the baKgatla in the Pilanesberg to the east,[71] and bears some resemblance to the Engers incident. Then, some English traders such as William Southwood found that the war offered chances to further his already established reputation for venality; firstly, to sell arms to the baHurutshe in their reserve and, secondly, to acquire a contract for the sale of grain to the British garrison in Mafikeng.[72]

On the whole, the war brought a halt to certain developments in the Marico district. Productive activities on white-owned farms declined considerably. The suspension of farming operations, and rent obligations, stock losses and rising land prices made a return to the land an unattractive prospect.[73] It took several years of determined state intervention before white agriculture was to revive and former exploitative relationships were resumed – which offered a breathing space for rural communities, the baHurutshe and the baKgatla, to gain a hold over the productive resources of land and cattle. After the war, relationships between the baHurutshe and their white neighbours never carried the same degree of violence and coercion that characterised pre-war relations. As we have noted, the mood of the rural Transvaal changed in the years immediately after the war when African peasants and African farm workers were reported to be recalcitrant and unwilling to accept forms of authority to which they had formerly been subjected.[74] Like the baKgatla in the Pilanesberg, the baHurutshe had also forged a close relationship with the British authorities by the war's end and might have expected to receive favourable treatment, a sentiment expressed in a petition to the secretary of the Law Department in 1902, in which the Hurutshe chiefs hoped for a 'continuance of the confidence established between our people and the ruling authorities'.[75]

The Tshidi baRolong in Mafikeng[76]

That we have firm evidence about baRolong involvement in the war is due to the publication in 1973 of Sol Plaatje's Boer War diary, the only known account of the famous siege of Mafikeng from an African perspective. On the eve of the South African War, the baRolong boo Ratshidi of the Mafikeng area of what was then the northern Cape sided with the British forces – essentially because of more than two decades of conflict with the Boers over land. From the late 1860s, the Boers in the Zeerust area of the South African Republic persisted in expanding westwards, usurping the ancestral lands of the Ratshidi. This resulted in several battles and Boer sieges of Mafikeng during the 1870s and 1880s, and in October 1899, when Boer forces under Commandant Piet Cronje were poised to invade Mafikeng, Rolong chiefs requested the commander of British forces in the area, Colonel Robert Baden-Powell, for arms and ammunition for their defence.

At first, the British prevaricated, but eventually the obvious danger posed by Cronje's forces changed this policy and Baden-Powell gave Snyder rifles and ammunition to about 400 baRolong men. The coloured community in Mafikeng also formed their own resistance contingent, the 'Cape Boys', while the Indians in the town were armed and incorporated into the white 'town guard'. Another two African groups, the Mfengu contingent and the 'black watch', were also given arms and ammunition. Each of these groups was assigned a specific portion of what is today greater Mafikeng to defend.

On 25 October 1899 the baRolong repulsed a Boer attack, killing an undetermined number of the attackers. This action encouraged Baden-Powell to give the baRolong defenders more rifles and ammunition and to increase the number under arms to about 500. On numerous occasions the baRolong, of their own volition, attacked the Boer positions on the outskirts of the town, sometimes capturing guns and equipment. The baRolong were nominally under the authority of Sergeant Sydney Abrams, but in practice they conducted their own operations of defence and resistance against the Boers, under their own command structure. Their commanders even used British-style military titles: Wessels Montshiwa was 'field-marshall'; Lekoko Marumolwa a 'general'; and others were sergeants and corporals. Blacks also generally acted as spies and dispatch runners, supplying the British military authorities in Mafikeng with much-needed intelligence about Boer commando activity. Those caught by the Boers, of course, ran the risk of being shot – and many of them were.

To conclude this section, we have noted the many similarities between the role and involvement of the baHurutshe in the Marico and the baKgatla in the Pilanesberg during the war. Both were willingly involved but also through the support and encouragement of the British military authorities. Both districts were characterised by a general lawlessness, including the looting of Boer properties (in particular cattle) with the baKgatla involved on a greater scale, inviting an interpretation (by Krikler) of the war as a 'peasant' revolt against a previously dominant agrarian ruling class. The end of the war saw the British administration ensuring a quick restoration of law and order and a return to pre-war master-servant relations, on the land, between the African people of the bushveld and local white producers.

ENDNOTES

1   P. Warwick and S.B. Spies, *The South African War*, pp. 186-209; Warwick, *Black People and the South African War*, pp. 45-46, 48-49. See also H.T. Siwundlha, 'White Ideologies and Non-European Participation in the Anglo-Boer War,

1899-1902', *Journal of Black Studies* (December 1984), pp. 223-234; R. F. Morton, 'Linchwe I and the Kgatla Campaign in the South African War, 1899-1902', *Journal of African History*, 26 (1985), pp. 169-191; W. Nasson, 'Doing Down their own Masters: Africans, Boers and Treason in the Cape Colony, 1899-1902', *Journal of Imperial and Commonwealth Studies*, 12, 1 (1983).

2   W. Nasson, 'Warriors without spears; Africans in the South African War, 189-1902,' *Social Dynamics*, 9, 1 (1983), pp. 91.

3   See, for example, the older explanations by J.S. Marais, *The Fall of Kruger's Republic* (Oxford: Oxford University Press, 1961); T. Pakenham, *The Boer War* (Johannesburg: Jonathan Ball, 1979); P. Warwick, *Black People and the South African War, 1899-1902* (Cambridge: Cambridge University Press, 1983). See also I. Smith, *The Origins of the South African War, 1899-1902* (Harlow: Longman 1996), pp. 70-97.

4   This section of the chapter draws upon an earlier chapter by one of the authors, Bernard Mbenga, 'The Role of the baKgatla of the Pilanesberg in the South African War'; G. Cuthbertson, A. Grundlingh and M-L. Suttie (eds), *Writing a Wider War: Rethinking Gender, Race and Identity in the South African War, 1899-1902* (Athens OH and Cape Town: Ohio University Press and David Philip, 2002), Chapter 5.

5   Morton, 'Linchwe I', p. 180.

6   T.S. Pilane, interview by Bernard Mbenga, Saulspoort, 29 May 1993; B.N.O. Pilane, interview and again, Saulspoort, 8 October 1993.

7   B.N.O. Pilane, interview.

8   Warwick, *Black people and the South African War*, p. 192-193.

9   L.S. Amery, *The Times History of the War in South Africa*, vol. III (Sampson Low, London, 1905), p. 200.

10   Ibid.

11   Amery, *The Times History of the War in South Africa*, vol. II p. 270. See also E.F. Knight, *South Africa After the War: A Narrative of Recent Travel* (London: Longman, 1903), p. 271.

12   R. Ramolope and S. Molope, joint interview, with Bernard Mbenga, Saulspoort, 17 February 1994.

13   A.K. Pilane, 'A note on episodes from the Boer war,' *Botswana Notes and Records*, 5 (1973), p. 131. This Pilane, an amateur historian and a member of the baKgatla royal family, was personally involved in the baKgatla-Boer war, but has since passed away.

14   Amery, *The Times History*, *vol. IV* (1906), p. 204. About the comparatively strong military position of the Boers at the beginning of the war, see pp. 270-271.

15   H. J. Botha, 'Die Moord op Derdepoort, 25 November 1899. Nie Blankes in Oorlogsdiens', *Militaria*, 1 (1969), p. 59.

16   J. Ellenberger, 'The Bechuanaland Protectorate and the Boer War,' *Rhodesiana*, XI (1964), p. 5.

17   Morton, 'Linchwe I', p. 179.

18   The captive Boer women and children were all repatriated to the Transvaal at the end of November 1899.

19   Botswana National Archives (BNA), Archive of the Resident Commissioner, (RC) 5/4, Affidavit of R. Pilane to Ellenberger, 2 August 1901.

20   BNA RC 5/4, Affidavit of Mongale to Ellenberger, 2 August 1901.

21   BNA RC 5/4, Affidavit of Mongale, 2 August 1901.

22 BNA RC 5/4, Affidavit of Lizzie de Villiers to Charles Bell, Resident Magistrate, Mafeking, 11 July 1900. In the same file, see also the affidavits of Segale, Ramono, Seroke, Thaperi and Seabatho (all of Mochudi) in which they admitted that they had stolen and shared the money.

23 BNA RC 5/4, T.H. Focke to Sir A. Milner, 12 February 1900.

24 M. Davitt, *The Boer Fight for Freedom* (London: Funk and Wagnalls, 1902), p. 174.

25 W. Fouche, *Pieter Stofberg: Zyn Leven, Arbeid en Afsterven*, cited in Schapera, *A Short History of the BaKgatla-baga-Kgafela of Bechuanaland Protectorate* (Cape Town: School of African Studies: Cape Town, 1942), p. 43. The quotation belongs to the Rev. P.B.J. Stofberg, a DRC missionary in Mochudi at the beginning of the war.

26 SNA 59 NA 2111/02, quoted in R. Williams, Resident Commissioner, Mafeking, to Sir G. Lagden, Johannesburg, 23 May 1902.

27 BNA RC 5/4, affidavit of De Villiers to C. Bell, Resident Magistrate, Mafeking, 11 July 1900.

28 For details, see J. H. Breytenbach, '*Ongepubliseerdemanuskrip*', typescript in Phuthadikobo Museum, Mochudi (April 1985), pp. 29-30, 34-35.

29 *Everdingen*, quoted in Schapera, *A Short History*, p. 42.

30 I. Schapera (ed.), *Ditirafalotsa Merafe ya Batswana* (Alice: Lovedale Press, 1940), p. 182.

31 I. Schapera, *Praise-Poems of Tswana Chiefs* (Oxford: Oxford University Press, 1965), pp. 98, 101.

32 See J. S. Mohlamme, 'The role of black people in the Boer Republics during and in the aftermath of the South African War of 1899 -1902,' Ph.D. thesis, University of Wisconsin-Madison (1985), pp. 40-41.

33 SNA 116 NA 672/03, SNC to NC, 27 April 1903, p. 2; Schapera, *Ditirafalo*, p. 182; *A Short History*, p. 20.

34 SNA 59 NA 211/02, cited in R. Williams to G. Lagden, 23 May 1902.

35 SNA 116 NA 672/03, SNC, the Pilanesberg, to NC, Rustenburg, 27 April 1903, p. 2.

36 Op. cit. p. 5.

37 TA , Van Warmelo Boxes K32/13, S.292(24), J. Masiangoako, 'Military system and war,' p. 2.

38 SNA 71 NA 2482/02, NC to SNA, 11 December 1902.

39 SNA 71 NA 2482/02, NC to SNA, 6 November 1902. In the same file, see also Telegram, SNA to NC, 23 May 1903; NC to SNA, 10 November 1903.

40 SNA 17 NA 396/02, Cited in R. Williams, R.C., Mafeking, to G.Y. Lagden, CNA, 14 February 1902.

41 SNA 116 NA 672/03, Petition by R. Pilane and others to Sir A. Lawley, 9 February 1903, pp. 3-4.

42 See, for example, Morton, 'Linchwe I," pp. 179-191; J. Krikler, *Revolution From Above, Rebellion from Below, The Agrarian Transvaal at the Turn of the Century* (Oxford: Clarendon, 1993), pp. 158-176.

43 Knight, *South Africa After the War*, p. 264 [our emphasis].

44 SNA 116 NA 672/03, SNC to NC, 27 April 1903, p. 3.

45 Op. cit., p. 2.

46 Knight, *South Africa After the War*, p. 270.

47 SNA 116 NA 672/03, SNC to NC, 27 April 1903, p. 3.

48 SNA 6 NA 519/02, Native Affairs Department, Annual Report for the Year Ended 30 June 1906, p. A.1.

49 TKP Vol. 239, Annual Report by the Commissioner for Native Affairs, Transvaal, 30 June 1903, Annexure D, p. B. 27.

50 Morton, Linchwe I, p. 188.

51 CAD, S.292(24) K32/13, Tsoabi, 'Shields', cited in J. Masiangoako, 'Military System and War', p. 6. It is likely that Tsoabi had witnessed the war. See also Morton, 'Linchwe I', p. 186.

52 For details, see Krikler, *Revolution From Above*, pp. 37-39. Quotation from p. 39.

53 SNA 169 NA 2059/03, NC to SNA, 18 September 1903, p. 35.

54 Krikler, *Revolution From Above*, p. 53.

55 A. Grundlingh, 'Protectors and friends of the people? The South African Constabulary in the Transvaal and Orange River Colony, 1900-1908,' in D.M. Anderson and D. Killingray (eds), *Policing the Empire: Government, Authority and Control, 1830-1980* (Manchester: Manchester University Press, 1991), p. 175.

56 SNA 71 NA 2482/02, High Commissioner to Lieutenant-Governor, n.d., January 1903.

57 106 NA 491/03, NC to SNA, 18 February 1903, p. 7.

58 BNA RC 6/13, L.K. Pilane, Mochudi, to Acting Assistant Commissioner, Gaborone, 26 November 1901. Most of the war losses, according to Linchwe, occurred around Saulspoort and Lesetlheng. In this source, see the 'List of property taken from the Bakhatla in the Transvaal ...'

59 Knight, *South Africa After the War*, p. 273.

60 SNA 71 NA 2482/02, High Commissioner to Lieutenant-Governor, n.d., January 1903.

61 Knight, *South Africa After the War*, p. 267.

62 TA Law Department, vol. 22, m. 952, Report by C. Levey, 20 October 1901.

63 *Hermannsburg Missionsblatt* (*HMB*), 1902, p. 99, Report from Dinokana.

64 TA, SNA, Vol. 48, No. 1644, Native Commissioner, Rustenburg, to SNA, 22 August 1902, Statement by Chief Gopane to E. Stubbs at Zeerust, 11 August 1902.

65 *HMB*, 1906, No. 1, p. 10.

66 *HMB*, 1902, pp. 198–199; *HMB*, 1900, p. 109. 4.

67 TA, LD, Vol. 24, No. 1885, C. Levey to Secretary, Law Department, 19 November 1900.

68 TA, SNA, Vol. 48, No. 1644, E. Stubbs to C. Griffith, memo, re: 'Looted Cattle in Possession of Chief Gopane', 13 August 1902.

69 TA, SNA, Vol. 48, No. 1694, Affidavit by Chief Gopane, 11 August 1902. In this document, Gopane stated that 'I have returned fourteen head of cattle to Mr Theneysen (*sic*) who resides near Zeerust (because) … Major Madoc (*sic*) advised me that Mr Theneysen (*sic*) had always been loyal to the British cause.'

70 TA, Magistrate's Reports, Marico, 5/1/1, No. 14, E. Stubbs to C. Griffith, 11 November 1902.

71 *Argiefvir Wes Transvaal Geskiedenis*, (Western Transvaal History Archive) Q. Postma Library, North West University, Potchefstroom Campus, J. H. E. Hamms Collection, pp. 11-13.

72 TA, SNA, No. 1125/02, Statement by A. J. Willis, 25 February 1904; W. Southwood to Baden-Powell, 28 June 1902.

73 See Krikler, *Revolution from Above*, pp. 13– 6. A very similar pattern emerged in the Orange Free State – see T. Keegan, *Rural Transformations in Industrialising South Africa, The Southern Highveld to 1914* (Johannesburg: Ravan, 1986), pp. 45-50.

74 Krikler, *Revolution from Above*, pp. 10-17. See also, B. Mbenga, 'The BaKgatla ba Kgafela in the Pilanesberg District of the Western Transvaal from 1899-1931' D. Lit et Phil thesis (University of South Africa, 1996), pp. 148–152; TDD, Vol. 239, *Annual Report of the Commissioner for Native Affairs, 1903*. In this Report, Sir Godfrey Lagden stated that 'during 1902-03 Africans had lost all touch with the Government and were practically out of control.' p. 2.

75 TA, SNA, Vol. 25, No. 772, petition forwarded by Secretary, Law Department, to SNA, March 1902.

76 This whole section on the participation of the Tshidi baRolong in the Siege of Mafikeng is based on B. Willan, 'Blacks in the Siege', I.R. Smith (ed.), The Siege of Mafeking, vol. 2 (Johannesburg: The Brenthurst Press, 2001), Chapter 6.

CHAPTER 3

Land, leaders and dissent 1900-1940

INTRODUCTION

During the years 1900 to 1940, many of the African chiefdoms in the bushveld were struck by dissension and what one historian, Graeme Simpson, has termed 'crises of control'.[1] The causes lay partly in past events, and partly through much closer involvement and engagement by these communities with the colonial world in which they lived and worked. More specifically, disputes over land acquisition and other material interests resulted in conflict, either among ruling factions or between the *dikgosi* and their followers, often united in opposition to the chieftainship in a 'popular' alliance. The other common feature of these conflicts was that they had a distinctly ethnic character as they were between opposing ethnic factions which mobilised support by appealing to ethnic sentiment and affiliation. Thus while ethnicity was the glue that kept the disparate factions of these communities together in the latter years of the nineteenth century, it increasingly became contested and a source of division in the first half of the twentieth. Nevertheless, despite internal dissent and division, all of the baTswana in the bushveld of the western Transvaal managed to adapt to changing circumstances. It was a difficult period of colonial intervention when chiefs were expected to cooperate with the state by implementing new administrative and bureaucratic tasks, but these black communities nevertheless managed to retain quite considerable economic and political independence and were not merely victims of the colonial state and the growing capitalist economy of South Africa.

What is also striking about these dissentions and divisions is that, in various forms, they prefigure the schisms and disharmony between commoners and traditional leaders and between different factions of chiefly families that struck many of the bushveld *merafe* in the early years of the twenty-first century, as the final chapter in this volume will relate.

The division of the baHurutshe

Moiloa, who had unified the *merafe* in the nineteenth century, was followed as chief of the Moiloa faction at Dinokana by Ikalafeng. When Ikalafeng died in 1893, his heir, Pogiso, was a minor, and Ikalafeng's brother Israel assumed control. In 1906 the government decided to instate Pogiso – but Israel opposed the appointment. The boy was only eleven years old, and Israel therefore felt he himself should continue as regent (such practices were common in Tswana societies and often the cause of schisms). The government was obliged to resolve the dispute, and upheld Pogiso's right to leadership. Indeed, the native affairs commissioner, Ernest Stubbs, was keen to see the back of Israel, who was considered 'unfit' for chieftainship on the grounds that he was a 'drunkard' and had been unwilling or unable to collect taxes shortly after the South African War.[2] It was probably the need to regulate the affairs of African rural communities such as the baHurutshe which took precedence over any considerations of 'native custom'. In April 1906, the Executive Council of the Transvaal approved Pogiso's appointment, believing that the majority of the *merafe* would support Pogiso and that Israel would quietly relinquish authority.

Israel, however, had significant support, and mobilised the non-Christian community behind him.[3] Moreover, Pogiso had few real roots in the reserve, having been baptised by the Wesleyans near Pretoria and then taken employment at a mine in Krugersdorp.[4] Israel refused to back down, and for six months in 1906 a chaotic situation existed in Dinokana, with Israel refusing to give up his duties and Pogiso trying to set up his authority.

Israel was determined not to give up his perceived right to rule over the baHurutshe at Dinokana. He went to Pretoria to see the secretary for native affairs, G. Lagden, and after meeting with no sympathy, he collected £150 from his supporters and went to Cape Town to enlist the services of Francis Peregrino, a West African who was editor of the *South African Spectator* and leader of the Coloured Peoples' Vigilance Association. Peregrino presented a petition to the commissioner for native affairs in the Transvaal,[5] pleading Israel's case. Shortly afterwards, Lord Selborne, recently appointed high commissioner, embarked on a tour of the Transvaal, intending to ascertain from rural Africans any possible causes of discontent. Israel's supporters saw their chance, turning up in their

FIGURE 7: Pogiso Moiloa when he took control at Dinokana. He died in October 1918.
Source: BaHurutshe Tribal Office, Dinokana

thousands in Zeerust to question Selborne about Israel's dismissal, and put forward their wishes about his future.[6]

Selborne was unmoved. This was not what he had travelled to Zeerust to discuss. The incident appeared to stiffen the resolve of the government, which ordered Israel out of the reserve in January 1907. But the ex-chief proved to be a thorn in the side of the state. He slipped quietly into Bechuanaland, and his key supporters stayed in the reserve, for which several of them were imprisoned for three months. The bulk of his followers moved onto two farms, Petrusdam and a portion of the farm Leeuwfontein which had been purchased by the baHurutshe in 1877. They were ordered off the farm (in fact the authorities seemed unaware that the chiefdom had bought the farm) but by January 1908 had still not moved. The Zeerust magistrate reported: 'The attitude that they have now taken up is absolute defiance to any authority.'[7]

The impasse dragged on throughout 1909. Israel moved from Bechuanaland to a farm near Barberton in the then eastern Transvaal. Towards the end of the year the position of Israel's followers on Leeuwfontein was much more secure as the purchase of the farm had now been confirmed, and they acquired another farm, Braklaagte, a short distance away, for a price of £1 500. Israel, shrewdly, did not disclose to Pogiso that he had collected money for the farm and registered it in his name (again, the fact of the farm's purchase only came to light a decade later).[8] In 1910, Israel joined his retinue at Leeuwfontein after three years of separation. Thus the Dinokana baHurutshe became divided, and relations between the two factions were strained for pretty much the rest of the century. The baHurutshe case is similar to those discussed below in the sense that it captures the difficult, confused and essentially fragile position of the state in these intra-African squabbles.[9] The baHurutshe troubles also reveal how land, privately owned, could become a basis for political autonomy.

The baFokeng – the challenge to August Mokgatle

The evidence for a decline in the popularity among the baFokeng of *kgosi* August Mokgatle (the grandson of Mokgatle Thethe mentioned in Chapter One) is visible in a series of challenges mounted against him in the first three decades of the twentieth century. These began in 1908 when the ownership of two farms, Turffontein 302 JQ and Klipfontein 300 JQ, was contested in the Supreme Court of the Transvaal. The baFokeng plaintiffs argued that the two properties had been purchased in 1878 by their forebears, and had been occupied by them ever since then – however, as purchases by Africans had to be registered in the name of a 'chief' with a specific 'tribally' designated following, the names of the original purchasers were not indicated on the title deed. The Court ruled that Africans could not hold land independently from the chiefdom as a whole, and found in favour of the defendants.[10] This had crucial ramifications for most of the land-owning bushveld *merafe* in years to come. In fact, the Fokeng *kgosi* at the time, Mokgatle Thethe, had made some contribution to the purchase of the properties, and the extent of the plaintiffs' autonomy from the chiefdom as a whole was not clearly established. The dispute caused ill-feeling between the plaintiffs and the ruling family, particularly as the former had to pay the costs of the case, which amounted to nearly £1 000.[11] The group who tried by this means to obtain individual tenure over the farms was from quite a wealthy and powerful section of baFokeng society. The political and ideological disputes which lay at the heart of the case were a hint of things to come.

In addition, there were other rumblings among the baFokeng which reflected growing divisions. The first related to a schism which had taken place in the 1890s, between the two sons of Mokgatle Thethe, Bloemhof and Tumahole, following the death of their father. At that time Bloemhof had been banished from Phokeng but some time around 1911 *kgosi* August Mokgatle allowed a number of Bloemhof's followers to return to the town. This did not imply that these 'rebels' (as they were later termed) had mended their differences with the then current leadership – in the dissent which was to follow they were prominent among those who caused the crisis. However, the schism was also reflected in material terms, specifically as an attempt to control the purchase and use of land and the rights of *dikgosi* to extract and disburse tribal levies and fines. It was thus a material and an ideological challenge to August's hereditary legitimacy and authority.

The 'rebels' alleged that between 1911 and 1916 August Mokgatle had misappropriated tribal levies by using them to pay off his personal expenses on a mill; by purchasing goods from a local store for his own consumption; and by retaining for himself the balances earned on certain levies earmarked 'for the fencing of the Phokeng reserve'.[12] Essentially, they were accusing him of corruption by using his office for personal gain. He was furthermore accused by the Fokeng *legotla* of being a 'drunk who was given to abusive language'.[13]

But the root cause of their anger was the purchase of the farm Welbekend which *Kgosi* August and his four sons had privately bought in 1919. The deal had been rejected by the secretary for native affairs, who insisted that the farm be registered in trust for the tribe, and the acquisition had also been opposed by a significant number of August's own *lekgotla*. August then proposed the idea of a £5 tribal levy (per person) to purchase the farm communally, but this was unacceptable to some because it would amount to more than the cost of the farm. Matters became all the more confused when 200 baFokeng men, many of whom had cattle posts at Welbekend, tried to buy the land themselves, bypassing August altogether. These developments had already created tension among the baFokeng when, in 1921, there was a furore over attempts to purchase another farm, Kookfontein 337. As a start, August Mokgatle had offered a price of £15 000 to the white seller, a certain J Kruger, a sum which was considerably more than the £8 000 agreed upon at a previous *lekgotla* meeting. In some circles, then, the *kgosi* was seen to be squandering tribal funds on a farm from which little general benefit would be derived. Kruger, whose asking price was £23 000, broke off negotiations at this point, forbidding the baFokeng to graze, water or keep cattle on the property, and ejecting a number of families living on it. The *lekgotla* then decided to boycott a mill and two stores on Kruger's farm. However, the boycott was not observed by August and the *lekgotla* promptly imposed a £10 fine on him – which was increased to £50 and finally £500 when that was not paid.

While opposition to August Mokgatle was gathering, the baFokeng were dealt another blow. The *lekgotla* changed its tune and agreed that Welbekend should be communally purchased in trust for the baFokeng tribe.[14] However, many people considered the price too high and resisted paying tribal levies, with the result that the payments were not made and the baFokeng were threatened with legal proceedings.

Simpson argues that the conflict of 1921-1924 was 'over access to and control over resources', with the *kgosi* was trying to maintain a measure of private control.[15] But it was not primarily a dispute between commoners and the *kgosi*. Among the 'rebels' were a significant number of councillors, headmen and followers of Mokgatle himself.[16] Moreover, they were led by David Mokgatle and Monnafela Mokgatle, both of whom were related to the ruling family and, for a long period of the friction, exercised almost total control over the baFokeng *lekgotla*. Another, Simon Mokgatle, was a cousin of August's, and Bartolomia Monnafela Mokgatle was his uncle. Yet another of the leading dissenters was Lucas Mokgatle, also 'of royal blood'.[17] This unusual aspect of the case was noticed by the *Star* newspaper: 'It is a remarkable fact,' it observed, 'that the rebel sections contain several members of "royal blood", as they claim to be the original BaFokeng, and they contend that the "loyalist" section of the tribe is another race.'[18] Even Simpson, who casts the dispute more in the mould of a challenge to the chief's authority to administer and control material resources within the chiefdom,[19] concedes that 'opposition to August Mokgatle found organisational and institutional expression in the *lekgotla* of Phokeng'[20]. With one exception, all members of the *lekgotla* were opposed to the chief, and it was these majority *lekgotla* members who were the leaders of the dissenting faction. Thus, 'it was through the legitimacy of the *lekgotla* that a challenge was posed to the chief's political power'.[21] The crisis in Phokeng in the 1920s, therefore, can also be seen as one between an autocratic leader and his inner circle of power holders as much as an intra-Fokeng contest between different factions for control of the material resources of the chiefdom. In this dispute the 'rebels' were supported by a large number of commoners or ethnic 'strangers' who regarded the rebel *lekgotla* and its representatives as allies whose interests merged with theirs in opposition to August Mokgatle.

Another view is not to regard the dispute in a material way at all, but as the consequence of a longstanding chieftainship squabble which could be interpreted as an attempt by Bloemhof's adherents to oust August Mokgatle on the grounds that he was not only unfit for leadership but was an 'illegitimate' ruler in any event. August Mokgatle's indiscretions and poor governance thus provided a timely opportunity for his opponents to mobilise opinion against him.

Whatever the case, the government now had to intervene, and decisively so. Initially, a number of officials, it appears, were quite sympathetic to the 'rebels' – or certainly to their grievances. Hatchard, the native commissioner, wrote in 1922: 'August is a particularly weak individual, and though he is the recognised hereditary chief, has completely lost his authority over the majority of the tribe resident at Phokeng.'[22] The state's first response to the crisis was to try and bring about some form of reconciliation, but it was not possible. The 'rebels' had thrown down the gauntlet and were not prepared to back off. In addition, they were reported as having adopted a threatening attitude to *Kgosi* August Mokgatle by 'blowing trumpets around [his] house, threatening and insulting [him] and [his] messengers and singing inflammatory songs in and around Phokeng'.[23]

By 1923 a situation of dual control prevailed in Phokeng. The 'rebels' led by David Mokgatle and Monnafela Mokgatle had full control of the alternative *lekgotla*, and held meetings without the presence of the *kgosi*. On his side, August had the support of a number of loyalists, who claimed that their chosen leader had the right to levy any number of fines and tributes and to utilise them as he saw fit, on the grounds of traditional and customary privilege. The 'rebels' (a characterisation to which they now strongly objected) then produced a litany of complaints against August, most of them to do with financial mismanagement or misappropriation of 'tribal' funds. But they also raised the issue of his autocratic leadership:

> The *lekgotla* of Phokeng … feel that our administrative functions in all tribal affairs have been overridden, disregarded and totally ignored by the chief, who had taken upon himself the autocratic control of the tribe, and who now refuses to listen to us or seek our advice.[24]

Thus by the middle of 1923 the government was less concerned about the rights and wrongs of the dispute than the need to maintain law and order around Phokeng. It was unacceptable that two baFokeng *lekgotla* were operating, and this alone was seen as a challenge to white authority. It was decided, therefore, to uphold August's authority 'owing to the unreasonable attitude of the malcontents'[25] (a more obvious explanation was that the Department of Native Affairs depended on leaders such as August for the implementation of government policies and the administration of control). The government declared the 'rebel' *lekgotla* to be illegal, withdrew any recognition of it or of its decisions, and sent evidence of its course of action to the attorney-general for his comment and endorsement. However, the attorney-general, on a legal technicality (that August had at one point voluntarily submitted to the authority of the 'rebel' lekgotla) adjudged that no offence had been committed. The ruling was a blow for the Department of Native Affairs. The native commissioner for Rustenburg noted that the decision 'throughout the Union … undermined all attempts to exercise a system of tribal discipline or control'.[26] Possibly to offset this setback, the government therefore promulgated Section 4 of the Transvaal Location Orders which forbade the 'rebels' to gather without the permission of *Kgosi* August. Predictably, the 'rebels' ignored the regulation, and in September 1923 eight of them were charged with breaking it.

In a sense, the 'rebel' leaders had played into the hands of the government, exposing themselves to legislation which could be used to punish them. However, the case took a new turn when the dissidents employed two white advocates, Messrs Benson and Strange, to represent them. In the first charge against the accused, seven were found not guilty on the grounds of insufficient evidence that they had actually convened the various meetings that took place. Kefas Magano, the only one found guilty of breaking the location law, took his case to the Appellate Division where his appeal was upheld on the grounds that the location laws were *ultra vires*.[27] The decision of the court severely upset the government's plans and dented its authority over the administration of the African population of the Transvaal generally. In the view of the magistrate for Rustenburg:

> Local officials are now placed in an invidious position, as the upsetting of legislation which was intended to strengthen their hand in the carrying out of the Tribal Administration, has had the effect of making it considerably weaker. The discontented natives in every tribe throughout the country are 'au fait' with the situation and what their educated associates do not tell them, they learn from country attorneys, many of whom rely on practice derived from Native sources.[28]

These sentiments prompted a second and more vigorous attempt to bring the 'rebels' under control. In March 1924, the secretary for native affairs, Arthur Godley, established a commission of inquiry into the events at Phokeng which delivered a report chronicling their 'offences' and assigning responsibility for every breach of the law. Simpson suggests that the 'Godley report' indicated 'bias' in favour of August, and his 'failure even vaguely to understand the material grievances underlying the conflict'.[29]

Armed now with 'evidence' of criminal activities by the alleged 'rebels', the state intervened again in the later months of 1924. The actual trigger for this was the refusal by David Mokgatle and thirty-one of his supporters to pay tax, for which they were subsequently arrested, and the assault by Lucas Mokgatle on one of the *kgosi's* messengers – the first and only instance of open violence in the entire affair. Official thinking on the crisis now turned to the theme of deportation. Godley opined that 'the best method of dealing with these recalcitrant natives is to deport them to the Soutpansberg and locate them in the location originally set out for chief Malaboch [Mabhogo]'.[30] A removal order was eventually signed by the governor-general in October, and the nine ringleaders were ordered off any land belonging to the baFokeng. These men were Daniel Mokgatle, Lucas Mokgatle, David Mokgatle, Simon Mokgatle, Mmonafela Mokgatle, Solomon Magano, Kefas Magano, Johannes Molefe and Josiah Morobe.

Indications were that the African National Congress (ANC) had a role in the affair, as the authorities blamed the organisation for the challenge to August. 'To what may be termed the general Native Congress spirit which pervades a more enlightened section of the tribe, may be attributed all the trouble,' wrote Godley.

Support for the ANC among the rural populace of the western Transvaal had been strong in the hope that the organisation might reverse some of the clauses of the discriminatory 1913 Natives Land Act.[31] When this hope was not forthcoming the Communist Party of South Africa (CPSA) made significant inroads from 1925 among Africans in nearby Potchefstroom, and two leading lights of the party, Josie Mpama (also Palmer) and Moses Kotane, Mokwena wa Mmatlaku (who hailed from Pella near modern Swartruggens), were raising the profile of the CPSA in Potchefstroom and Rustenburg as was the Industrial and Commercial Workers' Union (ICU).[32] Nevertheless, Godley was probably reasonably correct in his assessment, for according to Limb, in a comprehensive study of the ANC's early years, the ANC 'maintained its structures in the countryside' and by the turn of the decade 'marched past forward past the ruins of the ICU' and the CPSA which had been hit by dissention in its ranks and state repression.[33]

By late November government statements were reaching a level of near hysteria over events in Phokeng, government's attitude fuelled by the increasing assertiveness of David Mokgatle and his supporters who revived their meetings, disrupted August Mokgatle's *pitsos*, and staged demonstrations in Phokeng. Thus the acting native sub-commissioner, Donald Hunt, described the situation in September 1924 as 'acute'. He went on to add: '… a considerable amount of blood was spilled at Bulhoek and I should not be surprised if there is equally serious trouble in Phokeng unless the matter is taken in hand and brought under control.'[34] In addition, there was a warning that Rustenburg's white population of 23 000 could become targets for attack.

A further dimension of the dispute between *Kgosi* Mokgatle and the 'rebels' was reflected in the tensions between the Lutheran Church and the Rev Kenneth Spooner's Pentecostal Holiness Church. Spooner was sent to South Africa in 1915 by his church in the United States. His arrival had immediately upset the Hermannsburg missionaries, who were already concerned about the disruptive influence of the Ethiopian churches in the Rustenburg district.[35] During the unrest, Spooner was singled out by the Department of Native Affairs for being the mastermind behind the dissent. Ernest Stubbs, the Rustenburg magistrate, accused Spooner of being 'in the employ of some revolutionary society, possibly red or communist and … at the bottom of the Phokeng dissensions'.[36] Another possibility, however, is that the Pentecostal Holiness Church 'shut out' support for the more radical African Episcopalian Church, and 'with it the ANC' in Phokeng.[37] Contrastingly, Penzhorn, the HMS missionary to the baFokeng, stood firm with August Mokgatle, so much so that he felt his life was in danger at the height of the unrest.

The degree of support for August in the closing months of 1924 is difficult to determine, given the bias of official correspondence. Open support for the 'rebels' was waning, but this could be attributed to the intimidating effects of state intervention rather than a sense of renewed loyalty to their *kgosi*. Whatever the case, this was certainly not the end of the affair. The nine appealed in the Appellate Division of the Supreme Court against their removal order. This failed, and they took the case on a second appeal to the court in Bloemfontein. It was dismissed again (with costs) in April 1925. The court was forced 'to assess the legitimacy of autocratic chiefly powers within African customary law as well as the relationship between the chief and his *lekgotla*'.[38] The need to uphold the authority of the *dikgosi* and 'tribal practice' thus became the cornerstone of the state's case, the weight of evidence finally convincing the learned judges Curlewis and Kindall to dismiss the appeal. A celebrated witness for the plaintiffs was the author and former chairman of the South African National Native Congress, Solomon Plaatje, who made a plea for the 'democratisation of chiefly authority'. After the judgement, in June 1925, the nine men and about 350 followers left Phokeng to live on the farm Witfontein, some fifty kilometres distant. Mokgatle forbade the deportees from 'breaking down their houses' (in order to reconstruct them elsewhere later) a petty act of vengeance for which he was criticised by the native commissioner. 'I advise you,' he wrote, 'to act like a father and to allow these people to take their houses, whether it is in the law or not.'[39]

The resistance against chiefly power was now broken. In Simpson's view, the event represented a clumsy attempt by the Department of Native Affairs to bolster the apparently dwindling popularity of chiefs in the district, which only 'exacerbated the developing crisis of chiefly legitimacy within the chiefdoms of the Rustenburg district'.[40] Developments among the baKwena ba Mogopa (discussed below) would certainly tend to confirm this view. Though ultimately buttressed by state intervention, August's reputation took a knock, and the strain of the affair, coupled with his continued abuse of alcohol, apparently took its toll, as in May 1927 he was reported to be 'wandering about beyond control', and was sent to Pretoria as a mental patient.[41] The baFokeng agreed to pay the costs of his hospitalisation, and after two months he recovered sufficiently to return home but although he was much improved his health never fully recovered. In 1930, David Mokgatle and six others made a half-hearted attempt to lift the expulsion order, hiring a lawyer, a Mr G Hutchinson, to petition the state on their behalf – subsequently, they contradicted their intentions by denying 'any desire to return to chief Mokgatle's rule'. From his side, Mokgatle refuted any suggestion that he was open to the idea of their return, describing Hutchinson as a 'madman'. Nevertheless, a number of the families who had left voluntarily in 1925 did subsequently return after expressing due regret and requesting forgiveness. Among them was Kefas Magano, who returned to Phokeng in 1933.

At the beginning of 1938, *Kgosi* August Mokgatle fell ill, and his eldest son, James Manotshe, a former student at Lovedale College in the Cape, was appointed

THE REVEREND KENNETH SPOONER

Kenneth Egerton Mosely Spooner was born on 8 January 1884 on one of the Barbados Islands, then a part of the British colony of the West Indies. He emigrated to the United States in his late teens and trained as a missionary of the International Pentecostal Holiness Church. Before departure for South Africa as a missionary, Spooner married an African-American woman, Malista Geraldine Warner. The couple arrived in Johannesburg via Cape Town in January 1915 and soon Spooner was sent to work among the baFokeng in Phokeng, just west of the town of Rustenburg, where he found the German Lutheran missionaries of the Hermannsburg Mission Society (HMS) who had been working among the community since 1866. Despite initial Fokeng opposition to his church, it quickly took root in the face of stiff and often hostile competition from the much longer established German missionaries – however, in Spooner's school the medium of instruction was English (unlike the Setswana of the German school), and it became one of the major attractions among the baFokeng, as did his introduction of sports, and sports competition with neighbouring schools. Many Fokeng parents withdrew their children from the Lutheran school and enrolled them in Spooner's, and Penzhorn, the HMS missionary to the baFokeng, forlornly lamented in 1917 that the 'deserters ... boldly declare that their children were learning nothing in my school, my school has not yet produced a clerk'.[1] Yet another attraction of Spooner's church was that, unlike the Lutherans who held baptisms inside the church, he held them in the open rivers, lakes and ponds, just as Jesus had done. This practice was new, and seemed fascinating. Spooner died and was buried in Phokeng in 1937. He spread the gospel and Western education far beyond the baFokeng community; by his death his church had expanded into many other parts of South Africa, resulting in sixty mission stations, seven primary and twenty-three high schools.

NOTES

1   *HMB*, no.12, Report for December 1917, p. 79.

FIGURE 8: Ethiopian baptism, near Rustenburg c. 1920
Source: National Archives of South Africa, Pretoria

to act for him. Mokgatle died three months later, after a 'long and painful illness' related to tuberculosis. He was seventy-one years old and had ruled for forty-one years, a feat for in no small measure he owed to the determination of many (but not all) of the officials of the Department of Native Affairs who bolstered his faltering status in Phokeng from the mid-1920s.

Crisis among the baKwena ba Mogopa

By the turn of the twentieth century the baKwena ba Mogopa was a divided *morafe*. It consisted of five sections, the most significant at Bethanie in the Rustenburg district and at Hebron and Jericho in the Hammanskraal magistracy, north of Pretoria. They were all within about fifty kilometres of each other geographically, and on land belonging to the Hermannsburg Mission Society (only later was Bethanie formally declared a 'native location'). At the beginning of the 1920s their recognised *kgosi* was Johannes Otto More Mamogale. He was not considered by the Department of Native Affairs or other commentators to be a particularly effective leader, but had the support of the *lekgotla* at Bethanie, and there was no outright opposition to him at the other two settlements. This was not to be the case for long, however, as the Hebron and Jericho factions attempted to assert their independence from him. The crisis among the baKwena was also reflected in an internal schism among the Bethanie faction which occurred towards the end of the 1930s.

Tensions between the baKwena at Bethanie on the one hand, and at Jericho and Hebron on the other had existed from the late nineteenth century.[42] A hint of what was to come, however, was contained in a petition sent in 1927 from some 'headmen' in Jericho who complained to the prime minister, JBM Hertzog, about Mamogale's 'misrule and misgoverning', specifically relating to misappropriation and misuse of tribal funds. Another and more potent bone of contention and source of violence revolved around ownership and control of land. The farm Elandsfontein had been purchased for the Jericho section in the name of Mamogale as the *kgosi*. However, the residents of Jericho accused Mamogale of not consulting them as to the price (which they considered to be too high) and of misappropriating the levies paid by the Jericho purchasers for his own use. The Hebron section also objected to the purchase on the grounds that they were now forced to contribute, as part of the chiefdom, to a farm from which they would derive no benefit. Similar objections were expressed by the Bethanie community. As Simpson notes: 'Implicit in the demands of the various groupings was the assumption that each section was responsible for purchasing its own land.'[43] Moreover, it raised a set of legal issues over in whose name land was to be registered, and who had exclusive access to it. Compounding the problem was the inability of the baKwena to actually raise the required funds to meet the instalments on the farm – and this rendered their other lands open to confiscation if they did not pay the outstanding debt.

The Department of Native Affairs, by its own admission, had little idea and fewer guidelines as to how to resolve the tangled web of related problems arising from the purchase of Elandsfontein. Prompted by the sub-native commissioner for Rustenburg, Mamogale called a meeting of councillors from all sections of the chiefdom which authorised the imposition of a £6 per annum levy on all adult males of the baKwena. This tactic failed, as it was simply ignored by many of the *morafe*. Mamogale had neither the practical capacity to root out defaulters and punish them nor the legal authority to resort to civil proceedings against them. The fundamental problem was that each section of the baKwena considered itself responsible and liable for contributing to the purchase of farms to which they had sole access and ownership. The concept of communal responsibility was giving way to a more individualistic and materialistic conception of land usage and the financial benefits to be derived from land.

Complicating the picture even further was the fact that the baKwena-ba-Mogopa had exchanged some of their land for a tract of land originally set aside for the Hartebeestpoort irrigation scheme. This had been done in accordance with the 1913 Natives Land Act, and was an attempt to allocate and consolidate land according to prescribed racial demographics. In whose name should this land now be registered? Their missionary, Reverend Behrens, argued strongly that the land transferred to the government had been purchased by the Bethanie section which should be the sole beneficiary of the land exchange. The HMS, after years of legal wrangling, was poised to transfer the farms of Jericho and Hebron to their Tswana converts; this raised the issue of whether they should be registered in the name of the minister for native affairs (as was the legal stipulation) in favour of the whole of the chiefdom, or only the beneficiaries concerned.

Despite the existence of two legal precedents laid down by the Transvaal Supreme Court in 1908, and later by its Appellate Division, which ruled that sections of 'native tribes' remained integral to the *morafe* as a whole, the secretary of native affairs suggested that the best way out of the impasse was to constitute the different sections as 'tribal entities' and that the farms be registered in trust for each respective entity. This in a sense was based on the reality that for practical and administrative purposes the different segments were independent from each other, and that Mamogale was incapable of bringing the Jericho and Hebron factions under his control.

Therefore, in March 1927 the Department of Native Affairs embarked upon a plan to formally divide the baKwena ba Mogopa. The sub-native commissioner responsible for Hebron and Jericho, J Hemsworth, called a meeting and asked the community to elect representatives to a meeting to be held at Bethanie and which was intended to gain the support of the communities for the envisaged division. Probably to his surprise, the idea was not met initially with much enthusiasm. The nature of the comments by the elders from Jericho and Hebron showed that they were suspicious of any such move coming from the government – they told Hemsworth that they did not necessarily oppose the concept of communal or customary rule through the *dikgosi*, but objected to the way in which Mamogale was abusing this system. Their main concern appeared to be that they should not be liable for levies for farms from which they derived no benefit. The 'passive resistance' of the two communities 'thus continued to express itself through widespread refusal to contribute to tribal levies'.[44]

Another mini-crisis arose soon after this failure to resolve the problem by separating the two baKwena factions. The lawyers of the white owner of Elandsfontein, John Reid, threatened to excise tribal land of the baKwena because of the continued failure of the chiefdom to pay off the remaining instalments on the farm. The Department of Native Affairs responded by raising a loan to meet the outstanding amount, the result of which was to 'replace one creditor for another [actually at a higher rate of interest] and to buy time for the Bakwena ba Magopa'.[45]

Having failed to get support for a division of the baKwena among the intended beneficiaries themselves, the state fell back on the only other recourse, which was to follow the view of the Supreme Court, as expressed in its 1908 ruling, that no segment of a tribal entity could hold land independently whilst remaining a part of a greater chiefly entity. From this point on the state remained steadfast in this approach, and attempted to strongly enforce it – but the state's intentions were sharply rebuffed when, in March 1925, the Hebron residents took John Reid and *Kgosi* Mamogale to court to have the sale of Elandsfontein set aside. The basis of the challenge mounted by the plaintiffs was that *Kgosi* Mamogale had failed to consult the entire *morafe* when he undertook to purchase Elandsfontein as he had consulted the Jericho and Bethanie factions only and therefore the contract was not binding on the Hebron section of the baKwena. Mamogale's predictable response was that as supreme ruler of the *morafe* he was not obliged to summon a general *pitso*, and he alone possessed the authority to enter into the sale agreement. The issue then hinged around the application of customary law, as had been the case with the baFokeng.

Although the case bore a close similarity to that of the baFokeng of Phokeng, the Department of Native Affairs handled it quite differently and led evidence both for and against the plaintiffs. For example, Ernest Stubbs, former native commissioner for the Marico district, argued that although formal land acquisition among Africans was a relatively new phenomenon, he believed that the consent of the entire community was necessary. 'Purchase of the farms without a *pitso* being called, in my opinion, would be contrary to native law and custom,'[46] he remarked. The sub-native commissioners for Hebron and Jericho backed Stubbs, adding that the baKwena ba Mogopa, a predominantly Christian community, had adopted the principles of democracy to a greater extent than non-Christians (a strange twist of logic based on the presumption that Christianity was inherently democratic) and that because they were Christians they were thus less likely to accept being ruled by an autocratic *kgosi*.

There were some differences within the Department of Native Affairs, however, over the extent to which Mamogale should have consulted with the entire chiefdom – whether he should have called a general *pitso*, or whether merely calling separate meetings of the various *lekgotlas* together with the *kgosana* (headmen) provided sufficient consultation. According to several of the Kwena witnesses, both procedures should have been followed, except that a general *pitso* of each section of the baKwena, rather than the entire chiefdom, would have sufficed. Among the witnesses called by the Hebron defence team were the historian and medical doctor Modiri Molema, and (again) Solomon Plaatje, who both argued that the consent of the entire chiefdom, through the holding of a *pitso*, was essential in such a crucial question as the purchase of land. Their presence as expert witnesses at the case reveals the extent to which both men maintained a connection with the rural population of the western Transvaal, and their desire to propagate 'democratic' principles in African customary law.[47]

The lawyers for the defendants similarly called a number of expert witnesses to challenge the Hebron plaintiffs. Among them was John Leary, a former magistrate in the Transvaal. Unsurprisingly, the thrust of the defence's case rested on the assertions of Leary and others that the *kgosi* had full autocratic powers, and that the *lekgotla* was merely an advisory body. Similarly it was suggested that the *pitso* was called by the *kgosi* to inform the chiefdom over a decision rather than to solicit its approval. As Leary argued: 'It is not the majority who prevail; it is the chief's will.'[48] Mamogale himself put it even more starkly. 'I notify them [the members of the baKwena ba Mogopa], I do not ask them.'[49] Another witness, Donald Hunt, the sub-native commissioner for Rustenburg, denied that the different factions of the bakwena ba Mogopa were 'distinct tribal entities' and that land, regardless of whose name in which it was registered, was 'tribal property'.

The witnesses for the defence provided an alternative interpretation of customary law that bestowed on Mamogale exclusive control over the acquisition and control of land as well as political control of the entire community. The problem was that these arguments were largely assertions and had not been written down, and thus had no proven prescriptive legal basis. To some extent, they were based on a romanticised view of customary law. At another level, they were a consequence of the system of trusteeship introduced by the colonial authorities after union to simplify land acquisition in the western Transvaal. Land had to be registered in the name of a 'chief', and was held in trust for the community by the secretary for native affairs. The system was open to abuse in that it allowed some *dikgosi* to regard and treat land as if it was their personal resource, particularly if they were allowed to get away with it. Consequently, as Simpson has convincingly argued, this was a 'materially rooted challenge and aspiration, [which] found ideological expression in the contestation of customary rules and processes'.[50] It would be wrong, however, to conclude that chiefs only acquired land for personal benefit. In many cases purchases were made by individuals or family groups and occasionally chiefs made personal contributions or organised the collection of funds for these acquisitions.

The divided opinions, not only among the baKwena ba Mogopa but also the officials of the Department of Native Affairs, made it a complex case for the presiding Supreme Court judge, Justice de Waal, who bemoaned the fact that the court should have to rule on what constituted 'native law' and what did not. He echoed Judge Curlewis's appeal in the baFokeng case for the codification of 'native law', as was the case in Natal, so that it should not be 'left to the unsatisfactory proof … of native evidence'. He concluded, however, that despite the baKwena ba Mogopa's long contact with the 'civilising' influence of Europeans (which in his view 'democratised' their form of government) the authority of the *kgosi* was supreme in all senses. Thus the purchase of Elandsfontein was deemed to be legal as Mamogale was considered to have consulted sufficiently with the *legotla* and, moreover, the governor-general had sanctioned the purchase.

The failure of the Hebron and Jericho factions to win their case, and the court's enforcement of the requirement to pay their levies, only served to stiffen their resolve to avoid any further payment and to secede from the larger political entity. This entailed a further change of tactics. The Hebron 'rebels' first took to petitioning the prime minister, JBM Hertzog, outlining the consequences of the Supreme Court decision – namely that several of their farms were in danger of being confiscated as a result of their liability for the debt on Elandsfontein – and they requested a commission of enquiry into Mamogale's running of baKwena affairs. Mamogale responded by challenging the status of the petitioners, claiming that they were not *bona fide* members of the Hebron *lekgotla*, and that their financial woes were a consequence of having to meet the costs of their continued litigation. Members of the Jericho section also petitioned the sub-native commissioner through their attorney. Their argument was that they were the experts in matter of 'tribal law and custom'. Not surprisingly, the Department of Native Affairs, on the basis of Judge de Waal's ruling, had little sympathy for the grievances of the petitioners, writing them off as an insubordinate minority.

The situation on the ground, however, contrasted strongly with the legal niceties. Most of the residents of Hebron and Jericho continued to resist payment of the levy and the result was that the baKwena ba Mogopa became increasingly indebted to payment for Elandsfontein and another farm, Nooitdgedacht. This resistance assumed what Simpson calls 'startling forms'[51] with many individuals refusing to recognise Mamogale on the grounds that they were 'detribalised' and had a right to secede from his rule. They also demanded the division of the immovable property and the debt of the *morafe*. At Hebron, the secessionists demanded the removal of Abraham Mamogale, the delegated 'sub-chief' of JOM Mamogale, and the 'rebel' *lekgotla* continued to meet in defiance of Mamogale. It was also reported that members of the *morafe* living in the urban areas had 'obstructed' the collection of levies, indicating the connectedness of the interests of migrants and rural dwellers. The Department of Native Affairs was reluctant to intervene as it might appear as if it was necessary to prop up Mamogale's waning authority – by this time the Department's officials were openly referring to the 'disintegration of the Bakwena tribe'.[52]

In Simpson's view, the challenges to chiefly authority reflected in the disputes among the baFokeng and the baKwena ba Mogopa, were indicative of a growing material prosperity among the African inhabitants of the bushveld. On one level, this opposition took on an ideological form in which the dissidents strove to introduce a more democratic framework to 'tribal' affairs. The determination of their *dikgosi* on the other hand, to hang onto and even entrench traditional and authoritarian forms of governance, only increased these tensions and widened the rifts between them. Yet it should be remembered that internal opposition to Batswana *dikgosi* was commonplace and occurred in contexts that had nothing to do with the emergent colonial order, so to interpret opposition as unpopularity is to set up something of a straw man. The state authorities, as the third set of actors in these events, were torn between the ultimately divergent aims of introducing 'civilised' government on the one hand, and ruling through the application of customary law on the other. The result was that the government basically muddled its way through these crises, and the colonial legal authorities, the judges of the highest courts in the land, were unable to assist in any fundamental manner.

The division of the baKubung

In an earlier chapter we traced how the baKubung had divided into two factions. One was based on the farm Elandsfontein or Molotestad, in the Boons district east of Ventersdorp, on the fringes of the bushveld, while the other under Lesele had been forced to leave the farm in 1886. At Molotestad disharmony was a constant feature until the end of the century and beyond. On assuming control in Molotestad, Ratheo Monnakgotla decided to expel the Anglican missionary, which proved to be an unpopular move leading to the formation of a pro-Anglican faction led by Ratheo's younger brother Elias. Ratheo subsequently invited in the Lutherans (the HMS), and a missionary, Dr Gervers, was stationed at Molotestad from 1893. Ratheo himself, however, refused to be formally converted. The Lutheran/Anglican division took on a material dimension over the control of irrigable lands (essential for the growing of winter wheat) which formed a portion of the farm (the significance of this resource is indicated by the fact that the baKubung realised a profit of £200 through the sale of wheat in 1902). Although the details are scant, the denominational quarrel was resolved with the expulsion of Elias and a number of his followers, including a member of the ruling family, Joel. Thus yet another baKubung faction departed from Molotestad.

Ratheo died in 1897, and his death and the uncertainties caused by the South African War prompted Elias to ask for 'forgiveness' from his regent successor, Abel.[53] This was granted, and he and his party returned to the farm. The rightful *kgosi*, Solomon, assumed office in 1902, and ruled until his death in 1918. Although no serious conflicts arose over the next decade, the schism in the community never quite disappeared. Joel took over as the leading figure representing Anglicanism, and this denominational discord found political form in Joel's apparent refusal to accept Solomon's complete authority. On one occasion he was sentenced to flogging by Solomon's council – an indignity he avoided by fleeing Molotestad. Although Ratheo had committed the chieftaincy to supporting the HMS, Solomon disputed the 'memorandum of agreement' he had entered into with the missionary society because it ceded extraordinary rights to the HMS over the Kubung farms Palmietkuil and Elandsfontein. In 1903 the British authorities concurred with Solomon, and ruled that Ratheo had no right to enter into the agreement as such powers were vested only in the state. The contract was therefore declared invalid, providing a good example of the government's custodial attitude in respect to African affairs,[54] and souring relations between the Lutherans and the community.

The Anglican/Lutheran divide lingered on. In 1931, when the congregation of the Anglican Church at Molotestad applied to build a new church on a public site, the *kgosi*, Richard Monnakgotla, refused to give his consent. The Anglicans decided to erect the church on a private plot belonging to Joel Monnakgotla. Richard immediately contacted the native commissioner in Ventersdorp to complain that the building was being constructed on 'tribal land'. The secretary for native affairs advised the commissioner that 'neither Joel Monnakgotla nor any other tribesman has the right to grant the church permission to erect a church on that portion of the tribal land allotted to him for his personal occupation'.[55] Nevertheless, he also expressed the view that the *kgosi* could not arbitrarily refuse permission for an Anglican church to be built at Molotestad, and recommended that the issue be placed before the *kgotla* to adjudicate and arrive at an agreement that suited all parties (it is not clear from the evidence whether the dispute proceeded beyond this point).

Harry Dugmore, a historian who conducted the first important research on the community, argues that the Anglican faction represented the messengers of new social and property relations, favouring privatisation (of land and its products) rather than communal ownership.[56] It followed that they were in the forefront of opposition to Solomon's continued demands for tribute labour.

Joel appears to have patched up relations with Solomon, but when the latter died in 1918, having succumbed to the influenza epidemic of that year, his followers took the opportunity to revive a number of old disputes with his successor, Hezekiah, the regent for Solomon's young son, Richard. Unrelated to the denominational conflicts, the discontent was about control over the irrigable lands, and charges of misappropriation of funds. The authorities confirmed that control of the 'land under water' was in the hands of the chieftainship, and that the charges relating to the improper use of tribal funds should fall away as Solomon, now deceased, was the only person who could provide the necessary evidence one way or the other. However, Joel and two others, Albert Rankoko and Amadeus Matsitse, were determined not to let the matter rest and issued a summons against Richard and the government. The matter went up to the Supreme Court. As was the case with the two communities discussed above, the judge upheld the right of the *kgosi*, Richard, to control access to and use of the irrigable lands.[57] The ruling had little effect upon Joel and his supporters, who continued the show their disapproval of Richard by refusing to make tribal contributions. Eventually, Joel showed his ultimate defiance of Richard by 'setting up his own tribal entity within [the] existing tribe',[58] and was forced by the authorities to leave Molotestad (this was about two years after the Supreme Court case in 1921). The incident may have indicated, as Dugmore argues, the state's desire to prop up the institution of chieftainship in the face of a growing ethos of individualism within Tswana *merafe*.

During these years the baKubung under Lesele lived in several different places, and may not have retained the same composition as the group that had departed Molotestad. According to informants interviewed by Dugmore in the 1980s, they lived in seven localities in an approximate radius of 200 kilometres of Molotestad; several of these places were located on white farms where the community lived as sharecroppers. In 1910, according to their traditions, Lesele's followers – who never numbered more than about thirty-eight families – saw in a newspaper that a farm belonging to a certain white man by the name of Coetzee was for sale for a sum of £2 800. Twenty-two family heads then approached him to arrange for the purchase. Their names appeared as registered owners, although according to an official source, fifty-three people contributed to the purchase. This was the core of the entity who in 1936 registered themselves as the baKubung ba Mathope, which gave rise to the name of their home, Mathopestad, in the Derby district of the western Transvaal.[59] The contributors were allocated land for use on the basis of their contributions; three of the wealthier families, the Rankokos, the Seboges and the Mosemes, were allocated the most land. The farm was a sound economic basis for the Mothopestad baKubung, who farmed maize, beans, sunflowers and sugar cane, and raised cattle and pigs, different families concentrating on different produce depending on their resources.[60] The residents remember selling a lot of their produce to local Indian and Jewish hawkers and traders from the 1920s to the 1950s. However the Reef was not a great distance away (about a hundred kilometres) and this enabled them to supplement earnings from agriculture with cash wages, especially in times of reduced production through poor conditions. Many migrants ended up staying on the Reef, and repatriating their wages to Mathopestad to provide support for the older generation and to educate their children.

The relative material wellbeing of the baKubung during these years prompted a number of families to try and privatise land, which led to conflict with those favouring communal tenure. This culminated in 1936 in a case between a commoner, Reuben Magano and Cyprian Mathope, the *kgosi*, who objected to Magano's fencing off of his land. After a lengthy hearing the baKubung appeal against Magano was struck off the court role, and Magano divided up his twenty-two acres of land into eleven portions, each of which was fenced and given to his family members and some supporters. The incident led to significant differences between what might be perceived as 'traditionalist' and 'progressive' (and mainly urban-based) factions within the community. Although this suggests, as Dugmore argues, that there was 'heightened competition for the finite resources of this community' (subsequent attempts to augment their landholding failed),[61] in the general context of conditions in the bushveld and western Transvaal the baKubung at Mathopestad seemed to have fared quite well. According to Dugmore's informants, relations between the baKubung ba Mathope and the baKubung ba Monnakgotla remained 'cordial' after 1910 when the former acquired their own place of residence, and marriages between members of the two groups were common. It would appear that these ties have progressively weakened over the years.[62]

The baKgatla ba Kgafela

The baKgatla, as we have seen, took full advantage of the South African War to enhance their economic position. They remained a divided community, with the Mochudi-based baKgatla in the Bechuanaland Protectorate controlling the affairs of those baKgatla based at Saulspoort (Moruleng) in the Pilanesberg. Broken promises about post-war reconstruction, coupled with growing fears that the Transvaal Crown Colony, especially under Louis Botha's prime ministership, might force the two factions to sever or significantly reduce contact with each other, prompted *Kgosi* Linchwe in Mochudi to embark on a land purchasing spree to ensure the material security of his Pilanesberg subjects.

Thus between 1906 and 1913 Linchwe bought 54.238 acres of land, increasing the baKgatla properties from 84.8 to 143.3 square miles.[63] Before the South African War the baKgatla had only managed to acquire four farms, and most lived as tenants on white-owned farms.[64] Linchwe died in 1924, and was succeeded by his son Isang, who took over as acting *kgosi* in that year (Linchwe's oldest son having died and his son Molefi being a minor). Isang continued to build upon this legacy of land acquisition. During his reign, which ended in 1929, he bought nine farms and began the process of buying two more, Kraalhoek 399 (acquired in 1931 and Welgeval 171 in 1933). In addition to buying land for their people, baKgatla chiefs also bought land which they registered in their own names. The late *Kgosi* Linchwe had set a precedent when, in 1877, he bought Holfontein 593, using his own and his people's resources. *Kgosi* Ramono, appointed as caretaker of the Pilanesberg baKgatla in 1909, followed Linchwe's example when, in 1911, he bought for himself and his family a one-eighth portion of the farm Doornpoort 251 and other properties. Even Isang who, despite being an acting *kgosi*, possessed all the powers and privileges of a permanent chief, bought property in his personal capacity.[65] Of the twenty-four farms the baKgatla bought in the Pilanesberg, eleven (or portions thereof) were registered in the names of Linchwe, Ramono, Ofentse (his successor at Pilanesberg from 1917) and Isang. Three of these *dikgosi's* farms, Witfontein 396, Middelkuil 8 and Cyferkuil 330, belonged to Linchwe alone. Two properties, portions of Koedoesfontein 42 and Doornpoort 57, were Ramono's. Ofentse owned one property; portion B of Welgewaagd 133, while Isang had five: Vogelstruskraal 400, Application 398, Nooitgedacht 406, Zwaartklip 405 and portion B of Varkensvlei 403. Altogether, these royals owned a good eighty-seven square kilometres of land between them[66] but the inference should not be drawn that these properties were for the sole benefit of the legal owners, as in many instances land was made available for occupation or use by members of the chiefdom.

On Ofentse's death in 1942 he was succeeded by his son Thari, who ruled only briefly, owing to ill-health. By that time, the late *Kgosi* Ramono's eldest son, Tidimane Samuel, had finished his studies at Tiger Kloof Educational Institution near Vryburg. In 1949, *Kgosi* Molefi, holding power in Mochudi at the time, appointed Tidimane, who had for a while worked as a clerk in the High Court in Lobatse, as chief in Moruleng. All the baKgatla-owned land in the Pilanesberg, including that owned by the chiefly class, constituted what was officially known as the Saulspoort location – the home, basically, of the community – where the baKgatla had their villages and cattle holdings and where they also developed infrastructure such as boreholes, dams and irrigation works. BaKgatla land in the Pilanesberg, however, suffered periodic droughts, and could not support all of its population – many, therefore, trekked to the nearby town of Rustenburg and the towns of the Rand for wage employment.

With regard to cattle ownership the Pilanesberg baKgatla did very well. Soon after the South African War, they had acquired some twenty-five cattle posts on farms along the Odi and Madikwe rivers that they had leased from white farmers. By 1910, Kgatla cattle herds in this area were described in the Transvaal Department of Native Affairs records as being 'immense' (the numbers of these herds had partly been due to the large-scale looting of Boer cattle from deserted farms during the South African War). *Kgosi* Linchwe was himself a major cattle holder, with many of his own posts along the border.[67] In 1922, during the reign of Isang, prospecting rights were given to Potgietersrust Platinum Ltd over seven of the Kgatla properties, following agreements between the two parties. In the late 1920s, platinum ores were discovered on two of Isang's properties, Zwartklip 405 and Nooitgedacht 406. The royalties paid for the prospecting boosted the coffers of the Pilanesberg baKgatla. Thus, from 1927, Isang received £625 annually as a royalty from this prospecting, on behalf of the Pilanesberg baKgatla.

Isang's rule was certainly very progressive in terms of looking after the affairs of his people, and he instituted a number of economic development projects in Mochudi and Saulspoort. However, even though Isang meant well for his people, the strict and even forceful manner in which he made them carry out such projects, also demanding levies from them, soon made him very unpopular. He also became embroiled in a conflict with Molefi when the latter became old enough to assume control. Although a succession issue, this also had a material dimension: in the mid-1930s, Isang was accused by Molefi and his mother of having stolen part of the estate of the late *Kgosi* Linchwe which was the inheritance of his descendants, including Molefi; and the main issue of the Molefi-Isang controversy centred on the farms bought in the Pilanesberg during Linchwe's reign. Isang's argument was that he and Linchwe had bought the farms in their personal capacities, and the properties ought therefore to be divided among all his sons, including Isang. But Molefi's counter-argument was that the farms, as well as Linchwe's cattle, were the properties of the whole *morafe*.[68] Molefi's argument was very favourably received by the *morafe*. Isang was further accused of having bought the support of the Pilanesberg baKgatla who settled on the farms he had acquired (a commission of enquiry into the dispute subsequently found Isang innocent of any wrongdoing). Land acquisition and control was a divisive issue here, much as it was among the other communities discussed in this chapter.

FIGURE 9 (left): Ramono Pilane, baKgatla chief in Saulspoort, 1903 to 1917.
Source: Mphebatho Museum, Moruleng (Saulspoort)

FIGURE 10 (right): Matlhodi, Ramono Pilane's second wife
Source: Mphebatho Museum, Moruleng (Saulspoort)

1   This phrase is taken from the title of a chapter from the commendable research of Graeme Simpson, from which much of the content of this chapter insofar as the baFokeng and baKhona are concerned is in turn drawn. See G. Simpson, 'Peasants and Politics in the Western Transvaal, 1920-1940.' M.A Thesis, University of the Witwatersrand, 1986.

2   National Archives (NA). All subsequent archival sources in this chapter are from this repository. Gov. Vol. 1090, 50/10/07, E. Stubbs to C. Griffith, NC Western Areas, 29 January 1906.

3   Manson, 'The Hurutshe in the Marico', p. 247.

4   Gov. 1090, 50/10/07, E. Stubbs to C. Griffith, 25 January 1906.

5   NTS vol. 324, r28/75, Acting sub-native commissioner, Zeerust to native commissioner, Western Districts, 29 January 1907.

6   *Rand Daily Mail*, 8 July 1907.

7   SNA, Vol. 394, no. 94, J.S. Leary to W. Windham (now secretary for native affairs), 8 January 1908.

8   NTS, vol. 324, r43/92, magistrate, Marico to secretary for native affairs, 18 January 1919.

9   For a summation see Manson, The Hurutshe in the Marico, p. 249-250.

10 SNA 83, 884/08 In the Supreme Court of the Transvaal, Modisakeng Petlele Plaintiff vs J.F. B Rissik and August Mokgatle, defendants.

11 See *HMB*, no. 12. December 1909, p. 387.

12 Native Affairs Files (NTS) no. 316, G. Godley to Secretary for Native Affairs, 15 March 1924, 'Enquiry into BaFokeng Dissention'.

13 NTS, 316, SNC to Native Commissioner, Rustenburg, 26 August 1923.

14 NTS 4351, 86/308, SNC to SNA, 9 June 1921.

15 Simpson, 'Peasants and Politics', p. 203.

16 Cited in Simpson, 'Peasants and Politics', p. 202, from NTS 3451, SNC to SNA, 16 August 1922.

17 SANA TPD 5/1925, para 11.

18 *The Star*, 14 December 1924.

19 Simpson, 'Peasants and Politics', p. 205.

20 Simpson, 'Peasants and Politics', p. 222.

21 Ibid.

22 NTS, 316, 15/55, NC, Rustenburg to SNA, 22 November 1922.

23 NTS, 316, Chief August Mokgatle to Sub-NC, Rustenburg, 21 February 1923.

24 NTS 316, 15/55, Phokeng *legotla* to NC, 26 August 1923.

25 Government Proclamation 760, 7 May 1923.

26 NTS 316, 15/55, NC to SNA, 22 November 1922.

27 Cited in NTS, 316, 15/55, Magistrate, Rustenburg, to SNA, 4 January 1924.

28 Ibid.

29 Simpson, 'Peasants and Politics', p. 220.

30 NTS, 316, 15/55, SNA to NC Rustenburg 5 August 1924.

31 For further discussion of the ANC in the western Transvaal see A. Manson and B. Mbenga, 'The African National Congress in the Western Transvaal/Northern Cape, c.1910-1964: Patterns of Diffusion and Support for Congress in a Rural Setting' *South African Historical Journal*, vol. 64, no 3 (September 2012).

32 J. Wells, *We Now Demand, The History of Women's Resistance to Pass Laws in South Africa* (Johannesburg: Wits University Press 1994), Chapter 5. Kotane only joined the CPSA in 1929.

33 P. Limb, *The ANC's Early Years: Nation, Class and Place in South Africa before 1940* (Pretoria: Unisa Press, 2010), p. 345.

34 NTS, 316, Hunt to NC Rustenburg, 11 September 1924. At Bulhoek, near Cradock in the eastern Cape, 163 followers of Enoch Mgijima, leader of a sect known as the Israelites, were killed defying government attempts to remove them.

35 Ethiopianism was an independent African church movement first established in the Eastern Cape in the 1880s.

36 Cited in Simpson, 'Peasants and Politics', p. 357, from JUS (Archives of the Dept. of Justice), 371, 3/1105/23, E. Stubbs, SNA, Confidential memo. 15 October 1924.

37 P. Limb, *The ANC's Early Years*, pp. 314-317.

38 This is Simpson's assessment. See 'Peasants and Politics', p. 229.

39 NTS, 316, 15/55, Native Commissioner, Rustenburg to Chief Mokgatle, 30 September 1925.

40 Simpson, "Peasants and Politics", p. 259.

41 NTS 316, 15/55, SNC Rustenburg to SNA, 23 April 1927.

42 These are discussed in Simpson, 'Peasants and Politics', Chapter 3.

43 Simpson, 'Peasants and Politics', p. 280.

44 Op. cit. p. 290.

45 Op. cit. p. 291.

46 Cited in Simpson, op. cit. p. 296 from NTS 325, 37/55, Rathibe and others versus J.O.M. Mamogale, p.28.

47 See Manson and Mbenga, 'The African National Congress in the western Transvaal/northern Cape', pp. 479-480.

48 Cited in Simpson, op. cit. p. 309from NTS 325, Rathibe and other versus Mamogale.

49 Simpson, op. cit. p. 313, Evidence of J.O.M. Mamogale.

50 Simpson, op. cit. p.307.

51 Op. cit., p. 317.

52 NTS 331 pp.55-56. NC to SNA, 27 February 1929.

53 Dugmore, 'The Bakubung of the Western Transvaal', pp. 60-65.

54 The correspondence relating to this dispute is contained in SNA, vol 84. No 2823.

55 NTS, vol. 699, r. 106/110, SNA to Native Commissioner, Ventersdorp, 6 May 1931.

56 Dugmore, 'The Bakubung of the Western Transvaal', p. 62.

57 NTS. 997/308, Record of evidence compiled by Chief Hezekiah Monnakgotla against Joel Monnakgotla, 24/3/1921.

58 Cited in Dugmore, 'The Bakubung of the Western Transvaal', p. 72 from N.T.S. 997/308 file 2. p. 4.

59 These developments are described in Dugmore, op.cit., pp. 83-85.

60 Op. cit., p. 86.

61 Op. cit., p. 94.

62 Interview with Esther Tshabalala, Elizabeth Lephoto, Josephine Moseneke and others, members of the baKubung royal family, 30 July 2012 by Andrew Manson and Bernard Mbenga. The informants denied the existence of any ties with the baKubung ba Mathope.

63   C.J. Makgala, *History of the BaKgatla baga Kgafela in Botswana and South Africa*, (Pretoria: Crink City Publications, 2009). pp. 167-119.

64   F. Morton, *When Rustling Became an Art, Pilane's Kgatla and the Transvaal Frontier*, 1820-1902 (Cape Town: David Philip, 2009), p. 279.

65   For details of BaKgatla chiefs buying and registering land in their personal capacities, see Mbenga, 'The BaKgatla -baga-Kgafela in the Pilanesberg District', p. 231.

66   Morton, *When Rustling Became an Art*, pp. 279-280.

67   Ibid.

68   Makgala, *History of the BaKgatla ba ga Kgafela*, pp. 233-234.

CHAPTER 4

'Away in the locations': Life in the Bechuanaland Reserves 1910-1958

Chapters Four and Five deal with the nature of life in the reserves of the former western Transvaal and northern Cape, which today form much of the North West Province. Reserves set aside for African occupation, were a common feature of the political and economic landscape of much of southern and central Africa. Initiated in Natal, the system was introduced by the British colonial administrations in other parts of the continent. But they were not only a British phenomenon: Moiloa's Reserve, as we have seen, was actually given to the baHurutshe by the Boers. The reserves shared a number of characteristics. Ideally, separate 'tribes' were meant to occupy distinct reserves; the land in them was allocated and used on a communal basis and it was generally inalienable – in other words, white farmers could not own land in the reserves. Legally, reserve-based Africans could technically not own land outside their designated areas, but the baHurutshe were able to expand onto farms bordering their reserve.

This account begins in 1910, the year of Union, and ends in 1955 with the enforcement of the Bantu Authorities Act which paved the way for the incorporation of the 'Bechuanaland Reserves' into the emerging bantustans. The reserves have possibly not received sufficent attention from historians and other scholars in relation to their importance, for they were home to millions of black people in South Africa, many being migrants who had families back home in the reserves. Hidden from sight, the inhabitants of the reserves or locations remained forgotten.

For many years the perceived wisdom was that in the twentieth century the reserves simply became reservoirs of labour for South Africa's expanding capitalist system, and that economic production in the reserves collapsed after about the first three decades of the century.[1] Such views were challenged subsequently by writers who recognised that the reserves (or former reserves) continued to offer a home to large segments of the rural population, who 'were not entirely undermined during the colonial era'.[2]

THE AFRICAN RESERVES IN SOUTHERN AFRICAN HISTORY

The historiography of the reserves of southern Africa begins with some of the leading figures in the liberal tradition, most notably WM Macmillan and C de Kiewiet, who gave considerable attention to poverty and landlessness in the South African countryside (and, in Macmillan's case, the cities).[1] However, southern Africa's reserves as a specific focus of study by historians[2] remained largely forgotten, even with the blossoming of Africanist historiography in the late 1960s. This was in contrast to a voluminous historiography on transitions on white-owned land in the countryside – perhaps not surprisingly, as most of the land in the country was by the mid-1950s under settler or state control. A turning point was reached in 1972 with the publication of Colin Bundy's seminal article (and later book) on the emergence and decline of a South African peasantry,[3] and the Palmer and Parsons volume *The Roots of Rural Poverty in South and Central Africa* in 1976. Bundy in particular provided evidence for the emergence of a market-orientated and innovative African peasant class where it had access to land. His work was the source of considerable debate on the nature of peasant society in South Africa, which itself was located within a wider literature regarding peasants and politics.[4] However, much of the ideological impetus for these works derived from underdevelopment theories, driven in the southern African context by Giovanni Arrighi and Harold Wolpe.[5] These scholars viewed the reserves 'largely in terms of their functionality to the developing capitalist system', that is, as labour reservoirs inevitably facing rural decline and dispossession'.[6]

This paradigm dominated the literature on the reserves. Certainly, aspects of Bundy's central thesis were critiqued. Jack Lewis, for example, questioned whether the migrant labour system was 'created mainly by forces external to the rural households' such as mining capital and the capitalist state and by Charles Simkins who pushed the decline of the reserves to at least the 1950s.[7] But the underlying assumptions, generally speaking, were not critically questioned until the early 1990s when the notion that the reserves were in a state of inexorable decline was strongly rebutted by Ian Phimister, who argued for the emergence of an 'entrepreneurial class' in the Southern Rhodesian reserves,[8] and by Barry Morton, whose work on the Ngamiland reserve in Botswana showed that the transitions in Ngamiland's economy between 1890 and 1966 'did not impoverish the region [and] in fact led to an improvement in the wealth of most of the population as well as a much more egalitarian distribution of wealth than had existed during the pre-colonial years'.[9] James Drummond and Andrew Manson stated the case for the continuance of commercial agriculture based on irrigation in the Hurutshe reserve, north of Zeerust (and a mere sixty kilometres from the Bechuanaland reserves), until well into the homeland period of Bophuthat-swana.[10] A very important book by William Beinart examined the transformations wrought by mercantile capitalism on Pondoland, in which he makes no reference to peasants at all, thus moving away from the tangled debates over the emergence of an identifiable peasantry, the influence of market forces upon it, or the chronology of its rise and fall.[11] Subsequently, from the mid-1990s, historical studies on reserve-based Africans all but dried up.

NOTES

1   W.M. Macmillan, *Bantu, Boer and Briton: The Making of the South African Native Problem*, (London: Faber and Gwyer, 1929; C. Kiewiet, *A History of South Africa: Social and Economic* (London: Oxford University Press, 1941).

2   South Africa's reserves were of significant interest to scholars of other disciplines. Hobart Houghton for example, conducted extensive work in the Ciskei. See D. H. Houghton and D. Philcox, 'Family Income and Expenditure in a Ciskei Native Reserve', *South African Journal of Economics*, 18, (1950), 418-438. Much of this research found its way into his classic work, *The South African Economy* (Cape Town: Oxford University Press, 1973). Also M. Horrell, *The African Homelands of South Africa* (Johannesburg: SAIRR, 1973).

3   C. Bundy, 'Emergence and Decline of a South African peasantry', *African Affairs* 71, 285, (1972) pp. 369-388. The full thesis was of course published as the *Rise and Fall of the South African Peasantry* (London: Heinemann, 1979).

4   For an overview see, F. Cooper, 'Peasants, Capitalists and Historians: A Review Article', *Journal of Southern African Studies*, 7, 2 (1981), pp. 248-314.

5   See G. Arrighi, *The Political Economy of Rhodesia* (The Hague: Mouton, 1967); H. Wolpe, 'Capitalism and Cheap Labour Power in South Africa: from Segregation to Apartheid, *Economy and Society*, 1, (1972), pp. 425-456.

6   W. Beinart and C. Bundy, *Hidden Struggles in Rural South Africa: Politics and Popular Movements in the Transkei and the Eastern Cape* (Johannesburg: Ravan, 1987).

7   J. Lewis, 'The Rise and Fall of the South African Peasantry: A critique', *Journal of Southern African Studies*, x1, 1, (1984) 24; C. Simkins, 'Agricultural Production in the African Reserves', *Journal of Southern African Studies*, 7, 2, (1981), pp. 256-282.

8   I. Phimister, 'Rethinking the Reserves: Southern Rhodesia's Land Husbandry Act Reviewed', *Journal of Southern African Studies*, 19, 2, (1993), pp. 225-239.

9   B. Morton, 'A Social and Economic History of a Southern African Reserve: Ngamiland, 1890-1966' (D. Phil. thesis, University of Indiana, 1996).

10   J. Drummond and A. Manson, 'The Rise and Demise of African Agricultural Production in Dinokana Village, Bophuthatswana', *Canadian Journal of African Studies*, 27, 3, (1993), pp. 462-479.

11   W. Beinart, *The Political Economy of Pondoland, 1860-1930* (Cambridge: Cambridge University Press, 1982).

We have not included in this discussion the fate of the baTlhaping living west of the Ghaap plateau in the Kuruman district. A 'socio-environmental' history of Kuruman has been written by the historian Nancy Jacobs. It charts how the baTlhaping on the Kuruman reserves adapted their lives and productive enterprises according to changing environmental conditions, and how this affected relations between them.[3] In addition to the fact that the Kuruman district has been 'covered', it is situated, strictly speaking, outside the North West Province, although its ecology and populace is the same as that of the Bechuanaland reserves, and some of the issues covered by Jacobs are therefore relevant to this account.

The Bechuanaland reserves were created in 1885, and formed part of the land defined later in the 1913 Land Act for black occupation, as was Moiloa's location. In that year, the British had sent an expeditionary force from the Cape to remove the two republics of Goshen and Stellaland, from the territory of the baRolong and baTlhaping respectively. The republics had been established after more than a decade of conflict between white freebooters (or mercenaries) and the baTswana who, up to that time, were nominally independent. The territory north of the colony of Griqualand West was declared the Crown Colony of Bechuanaland and was ruled from Britain. In 1895 it was proposed that the territory, including that north of the Ramatlabama spruit (that is, in modern Botswana), be placed under the control of the Cape Colony. Sensing colonial rule under the Cape would be harsher than being under direct British administration, the Tswana chiefs objected strongly and sent a delegation to Britain to state their case for continued British protection. The result was that a protectorate was declared over the territory north of Ramatlabama while the baTswana south of this became an extension of the Cape Colony. For the baTswana who still remained in the Cape, the position in the reserves set aside for African occupation was not altered fundamentally in the sense that land was inalienable and chiefs still retained quite extensive civil and legal rights over their followers, more so than in other parts of the Union of South Africa. However, the baTswana were still restricted to only eight per cent of the territory they had formerly occupied, most of which was confined to the immediate lands surrounding their main towns, thus cutting them off from more distant grazing lands to which they had formerly had access (see Map 6).[4]

Although the crown colony of Bechuanaland ceased to exist from 1896, the term 'Bechuanaland reserves' continued to be applied by administrators to refer to baTswana living south of the Molopo river towards Vryburg, Taung and Kuruman. The reserves can be divided into two. First were those in the Mafikeng district occupied mainly by the baRolong: the Molopo and Setlagole reserves (there was also a tiny reserve at Mosita) named after the two rivers that bisected them;

and 234 145 and 86 533 morgen in size respectively. Although it was the intention of colonial policy makers that reserves should encapsulate distinct 'tribes' or ethnic factions, this was not always possible – thus the Molopo reserve, inhabited mainly by the Ratshidi baRolong, had quite large minority ethnic factions within it, including the Rapulana and Ratlou with whom they had not, historically, enjoyed good relations, as well as enclaves of baTlharo and baTlhaping.

The baRolong boo raTlou boo Mariba, the baTlharo and the baTlhaping were allocated a jigsaw of eleven mostly smaller portions of land nearly all north-west of Vryburg itself. The baRolong boo raTlou boo Mariba under the chieftainship of the Letlhogile family were located in the Ganyesa reserve but also occupied the Jakkalsdraai, Tlakganeng and Morokweng reserves, the last being one of the largest (over 160 000 morgen) in Bechuanaland. Headmen were appointed by the Ganyesa chiefs to administer these reserves. Heuning Vlei was the 'capital' of the baTlharo, under the Bareki lineage. The baTlharo also occupied Madibeng and Dinopeng (misspelled Linopen) which were administered by headmen appointed at Heuning Vlei. Living on the Thakwaneng (misspelled Takwanen) reserve were the baTlhaping ba Phuduhutswana, a branch of the more senior chiefdom at Taung, who also occupied the Klein Chwaing reserve. The local native commissioner and magistrate resided in Vryburg (see Map 7).[5]

The baTswana living in the reserves faced a number of hardships. The land was in general not suited to agriculture; the average rainfall in the baRolong reserves was about 53 cm a year, while the south received only 38 cm. The quality of soils was generally poor, dominated in the south particularly by limestone outcrops, and the vegetation consisted of scrub grasses interspersed with kameelthorn and other varieties of thorn trees, which were probably more plentiful half a century ago than today. Droughts were frequent. The reserves were thus not suitable for arable production, and stock-keeping was the more important economic enterprise.

In addition, the area was often struck by an array of plagues and diseases. Locusts devastated the reserves in 1925 and again in 1933. To combat the locusts, government officials took to spraying the crops, but this in turn led to the death of many animals from arsenic poisoning.[6] Outbreaks of foot and mouth disease were another scourge. In 1933, in addition to the locusts there was a particularly virulent outbreak of foot and mouth and the Bechuanaland reserves were declared a quarantine area. Reverend Peasley, a missionary of the Society for the Propagation of the Gospel based at St John's mission at Moithitong in the Dinopeng reserve, drew the attention of the Vryburg native commissioner to the 'serious condition of the people of Linopen Reserve and the adjacent villages. Their crops have been entirely destroyed by drought and owing to the existing law about moving stock, they are unable to barter stock for food.'[7]

Prior to colonisation and the formation of the reserves, the baTswana had relied on being able to move animals to pasture in different areas, but this was now almost impossible. Moreover, in former years cattle were either obtained or traded in the Transvaal or the Bechuanaland Protectorate, but this practice was curtailed by colonial legislation. When the 1923 Native Affairs Commission visited Ganyesa, the commissioners were informed by *Kgosi* Letlhogile Ntselang that they had cattle posts in the Bechuanaland Protectorate and the restrictions were therefore a 'great hardship' and one of the reasons why they were starving and could not pay their taxes.[8] But the legislation remained in place.

As resources were limited for all people in the region, it led to increased competition. In the baTlhaping/baTlharo reserves this was a source of frequent conflict with the white community and there are many cases of whites illegally hunting, prospecting, grazing animals or quarrying in the reserves. One case, in 1925, led to an official investigation of a skirmish between *Kgosi* Bareki and some of his followers, and a group of white men from Vryburg, over illegal prospecting. The magistrate reported that '… it seems clear that for a number of years there has been friction between the baTlharo of Heuning Vlei and European hunters, traders and others visiting the reserve, and that the natives have adopted an insolent attitude towards white people'.[9] Despite this incident, Bareki was regarded by the magistrate as a 'man of good character' who had 'laboured under a sense of grievance against the government' and consequently 'has as little to do with government officials as he possibly can'.[10]

This situation was often reversed, with the residents of the reserve antagonising the white farmers by squatting illegally on their farms. The problematic situation was compounded by several factors. Firstly, the baTlhaping/baTlharo reserves were scattered about amid white farmland, some of which was unoccupied or occupied by (usually poor) white tenants who made little use of the available grazing. Secondly, the beacons indicating the boundaries of the reserves were often either torn down or missing in sections, so that the black occupants did not know they were living on white-owned farms. Thirdly, it was hard for the officials in Vryburg to resolve the squatting problem owing to the long distances involved in investigating the situation *in loco*, which prolonged and worsened the disputes. Competition leading to contests between the Tswana inhabitants of the various reserves was common, and the reasons generally more complex than white/black frictions. The problems were almost always expressed in material terms – over control of land, water sources or other economic resources – but they also had a strong political context. The antagonistic parties expressed their differences in terms of political rights and issues of leadership that had a longer historical context, as is amply illustrated by the following cases.

The Ratshidi baRolong and the Rapulana baRolong locked horns over control of a place called Lotlhakane (Rietfontein), a site about twenty-five kilometres south-east of Mafikeng. It was an important site, the location of a fountain or 'eye' which could irrigate the surrounding lands and made it possible to cultivate winter wheat which, from at least the 1870s, had been transported for sale in Kimberley.[11] The question of who were the rightful owners of Lotlhakane became the focal point for a long political tussle between the two.

In fact, there was a lot of enmity between the two Rolong chiefdoms, spanning the previous half-century or so. After the disturbances of the *difaqane*, the baRolong found themselves living with Wesleyan Missionaries at Thaba Nchu. In 1841 they trekked back to their former homeland in three groups: Ratshidi, Ratlou and Rapulana. The Ratshidi occupied land along the Molopo river, which was nominally unclaimed, while the other two chiefdoms occupied sites within the newly created Transvaal or South African Republic. Both Ratlou and Rapulana, however, had historical claims to the land where the Ratshidi had settled ahead of them. In addition, a large number of booRatlou lived at Setlagole and there was a Rapulana settlement at Lotlhakane itself. The divisions between the Ratshidi on the one hand and the Ratlou and Rapulana on the other widened as the Boers and the British fought for ultimate control of the region, the former siding with the British and the other two with the Boers. The issue of who owned Lotlhakane thus had greater implications than its productive potential. It was over who had political control of a wider region.

The close ties between the Ratshidi and the imperial power of Great Britain ensured that they enjoyed the upper hand in this contest for land and influence in what had now become the Bechuanaland reserves, and after Union in 1910 the Ratshidi felt they would be treated on equal terms. Against a backdrop of provocative actions by the Ratshidi *kgosi*, John Montshiwa, in 1913-1915, the two senior *dikgosi* of the Rapulana and Ratlou, Matlaba and Moshete, petitioned the secretary for native affairs in Pretoria to re-open the Lotlhakane case, one that had in fact been a matter of concern and investigation for thirty-three years by successive officials and representatives in the Bechuanaland region. Eventually, the Supreme Court ruled that the Rapulana (correctly) were the senior of the three factions and therefore Lotlhakane belonged to them.[12]

The ruling dented Ratshidi hopes, and the balance of political power definitely tilted more to the Rapulana/Ratlou, although the former remained important in the eyes of the colonial government and still retained a presence over Lotlhakane itself.

THE BAROLONG NATIONAL COUNCIL

An interesting aspect of the clash between the Ratshidi and Ratlou lies in the formation of a body known as the Barolong National Council (BNC). The BNC was formed
in 1915 but was a spent force by 1920. It was created at the height of the Ratshidi-
Rapulana conflict that was increasingly focussed on the ownership of Lotlhakane. The
first president of the BNC was the Reverend EM Sehumo, a member of the Rapulana
'royal house' and a minister in the African Catholic church, which had broken away
from the mainstream Roman Catholic Church in South Africa. The stated intention
of the BNC was to unify the baRolong through mobilising ethnic sentiment, but the
underlying objective, strongly suggested by the timing of its founding, was to dilute
the influence of the Ratshidi. As one commentator has stated, the BNC's formation
was 'politically motivated by the Rapulana and the Ratlou ... as a lever to elevate
Moshete as the paramount of the BaRolong'. The Ratshidi, unsurprisingly, had very
little to do with the BNC and saw it for what it largely was – a front to advance the
interests of their political opponents, particularly in Lotlhakane and the Setlagole
reserve. Eventually, the BNC imploded from within. In 1920, the Rapulana *kgosi*,
EM Matlaba, and Rev Sehumo were found guilty of embezzling £578 and £78 respectively from a fund they had established, in Matlaba's case to satisfy his well-known
fondness for liquor.

Conflict at Disaneng

A second challenge to Ratshidi power came from the west, from a Tlharo community under the Masibis living at Disaneng, about thirty kilometres from
Mafikeng. According to the Ratshidi version, the baTlharo had asked to settle
in the territory of the baRolong during the time of Montshiwa, who ruled from
1849 to 1896.[15] The two chiefdoms had enjoyed an amicable relationship during
the nineteenth century although the baTlharo paid tribute to the Ratshidi. But in
1910 this amity suddenly came to an end. The baTlharo now wished to exercise
independence from the baRolong, and objected to the appointment of a headman
by the then *kgosi* of the Ratshidi, Badirile Montshiwa. The matter went before
the circuit court in Mafikeng where the baTlharo argued that Montshiwa had
'invited them in' to settle in his country in about 1866. The judgement, however,
went against them, and the appointment of the headman was upheld[16] – and precipitated over three decades of bickering and political posturing, proving to be an
administrative nightmare for the government.

In 1931 Methusela (also Jan) Masibi was appointed chief at Disaneng following the death of his father and immediately instigated an aggressive policy against
Ratshidi overlordship. One of his first acts was to march with about sixty armed
men to the border with the intention, apparently, to 'assault chief Lotlamoreng

in the Protectorate' – the reason for this aggression was a rumour that Lotlamoreng had seized some cattle of Masibi's people in the baRolong Farms. Masibi instructed them not to obey Lotlamoreng's orders. The government responded to this 'disorderly conduct' by raising the possibility of sending Masibi back to the Kuruman district where most of the baTlharo were settled.[17] This seemed to pacify the Masibi group for a while, but in 1934 Lotlamoreng again complained about Masibi's aggressive attitude. In 1935 Lotlamoreng, possibly responding to Masibi's bellicosity, sent a party of men armed with guns, bayonets, axes and assegais to seize sixteen head of cattle belonging to the baTlharo. This same group of assailants then went on to assault a dozen of Masibi's subjects residing on the baRolong farms. Despite this, the baTlharo still refused to accept Lotlamoreng's nominee as headman at Disaneng, prompting him to appeal to the government to intervene in an act he described as being 'tantamount to open defiance of my prestige and authority'.[18]

Relations between the two parties had deteriorated to the point that in 1940 Masibi asked for permission to leave the district. However, this never happened and a compromise presumably was reached between the two parties whereby the baTlharo recognised Ratshidi authority, in return for which they were exempted from certain obligations – for example, they did not have to plough fields for the Ratshidi or pay towards the school levy fund. In essence it was more bluster than anything else on the part of the baTlharo, as they occupied some of the best land in the reserve.

Nevertheless, tensions flared up again in 1946. At issue was a strip of land running north to south between the designated areas of the baRolong and the baTlharo and claimed by both. Numerous followers of Lotlamoreng lived in it but paid no allegiance to Masibi. In turn, Masibi had ploughed land in the vicinity of Ratshidi cattle posts in this strip of land. The native commissioner for Mafikeng met with both men in February in an attempt to defuse the tension and find 'a way of disposing of the whole dispute between the baRolong and the baTlharo[19] but the boundary problem persisted until 1949, and was still not fully resolved by the mid-1950s.

At the core of the baTlharo-Ratshidi dispute were material issues. Tlharo lands lay alongside the Molopo river and boreholes could be, and were, drilled there during the 1930s and 40s, providing water for animals which were also located close to the disputed ploughed fields. Control of these two resources was critical in an environment where water and arable land were in short supply. These two disputes at Lotlhakane and Disaneng reduced the influence of the Ratshidi in the political affairs of the reserve, but only partially so, for they continued to enjoy a good relationship with the colonial administration.

The Lopurung affair

It was not only the Tshidi-baRolong who were faced with internal discord. At Phitsane, a Ratlou clan under Tshipitota Motseakhumo faced an internal revolt by the villagers of Lopurung. In 1950, the *lekgotla* at Phitsane resolved to remove the residents from Lopurung and resettle them at Phitsane. It was a resolution seemingly in compliance with so-called betterment plans to improve conditions in the African reserves, in this case to divide land for residential, agricultural and pastoral purposes.[20] The villagers, however, refused to move. This precipitated something of a crisis. Tshipitota threatened to demolish their residences, and identified five 'ringleaders of the rebel group'. This threat led to the headman, a man named Mosii, and 120 residents, approaching a legal firm in Rustenburg. Although the Department of Native Affairs stood right behind the *kgosi*, the Lopurung 'rebels' managed to get a temporary interdict restraining Tshipitota from moving them.[21] The native commissioner for Mafikeng advised the *kgosi* to appeal this decision but it took six years before the matter finally went on appeal to the Cape Division of the Supreme Court. Tshipitota's appeal was upheld. Shortly thereafter, Motseakhumo died and relations between Mosii and the new *kgosi* were much improved. This led to an 'adjustment' (presumably a compromise) between the new chief and the Lopurung people, and an amicable settlement.[22]

A number of interesting conclusions can be drawn from the Lopurung affair. The Lopurung 'rebels' were a fairly successful group of peasant producers. Mosii farmed on a two morgen plot and had fenced his property, for which he paid £150. He was reported to be selling cattle across the border in the Bechuanaland Protectorate. Additionally the residents of Lopurung 'grazed cattle towards the Railway Block [a group of Trust farms] for a distance of ten miles'[23] and if they were to be moved these grazing lands would be inaccessible. Then, it transpired that Mosii and Motseakhumo were competing to establish white traders in Lopurung from whom they could expect certain favours, in particular short-term credit.[24] The incident reveals the growing division between subject and ruler in cases where subjects depended less and less on chiefly support – and it shows how far the Lopurung residents were prepared to go to safeguard their interests. They were well aware of the advantages of enlisting the services of lawyers and were clearly able to meet the financial costs of fighting the matter as far as the Supreme Court. The resistance they mounted resembles that shown by factions of the baFokeng and baKwena ba Mogopa detailed in Chapter Three.

Production and economic enterprise in the reserves

Despite the generally unfavourable environment (which gave rise to increased competition and disputes), the inhabitants of the Bechuanaland reserves battled to overcome these hardships, and made progress, albeit modest and inconsistent progress. There was an increase in the numbers of cattle, sheep and goats belonging to the black occupants. In the Molopo/Setlagole reserves, for example, in 1937 there were just under 38 000 head of cattle, a number which grew consistently to 55 875 in 1949. Between 1944 and 1949 herds of sheep and goats increased from 43 962 to 62 866 and from 56 046 to 70 093 respectively. As Nancy Jacobs, writing about the Kuruman area, has remarked, goats 'reproduced quickly, required little capital and thrived on the bushes', and 'could be slaughtered more frequently than cattle'.[25] Agriculture was subject to the vagaries of drought and disease, but in good seasons the land could produce modest yields. The 1959 census indicated that the Ratlou at Setlagole had 1 558 non-irrigated morgen under cultivation which yielded 559 sacks of white mealies and 221 kilograms of sorghum. The Ratshidi (Mafikeng) had 3 031 morgen under cultivation yielding 994 mealie sacks and 311 kilograms of sorghum, and the Rapulana under Seatlholo cultivated 1 001 morgen yielding 652 sacks of mealies and 96 kilograms of sorghum.[26] Although these figures suggest that only about ten per cent of available land was under cultivation, the agricultural officer for Mafikeng concluded that 'the natives do not have to obtain excessive amounts of grain from other areas'.[27] Statistics for the baTlhaping /BaTlharo/boo Mariba reserves are hard to come by, but in the larger reserve of Morokweng, for example, it was estimated that there were 11 500 head of cattle and sheep in 1917-1918, a number that had grown to 13 000 by 1925.[28]

This measure of economic stability was made possible by increased attention to the provision of water supplies in the Bechuanaland reserves. In 1931, the minister for native affairs authorised an extensive programme for the sinking of boreholes to be carried out by the irrigation section of the Department of Agriculture. By the end of 1944 nearly forty boreholes had been sunk in the reserves around Mafikeng, and at least eighteen in those surrounding Vryburg. A further ten were installed near Phitsane by 1959. Many of the boreholes followed the dry river courses of the Molopo and Setlagole rivers and their tributaries.[29] In addition, many dams were built in the reserves – at least seventeen in the Molopo and Setlagole reserves alone. During the preparations for the borehole construction programme, the magistrate at Vryburg noted: 'If boreholes could be sunk these reserves could carry double the amount of stock and a correspondingly large number of people.' This proved to be absolutely correct, as the improved water supply afforded by the boreholes and dams allowed for increases in stock holdings. Thus the native commissioner at Vryburg, writing when the programme was well underway in 1936, noted: 'The availability of boreholes in most of the reserves has led to an improvement in the state of animals and a slight increase in numbers.'[30]

South Africa's Department of Native Affairs in the 1920s and through the middle decades of the century was concerned to make the African reserves as self-sustaining as possible, especially after the Great Depression years. It consequently sought to introduce progressive farming measures, some of which, as we have seen, proved unpopular. The Department of Agriculture had a staff of between eight and ten tasked with improving conditions in the Bechuanaland reserves; at least two were white agricultural specialists, but the rest were African demonstrators and rangers, men such as Morara Molema, E Gaboutloeloe and G Dlodlo, who were well versed in 'scientific' farming as they had studied at the Tsolo Agricultural College in the Transkei. They operated what were called

FIGURE 11: Morara Molema, brother of Dr Modiri Molema, was agricultural extension officer in the Bechuanaland and Moiloa reserves.

Source: Molema family

'demonstration plots' on which they introduced modern farming methods. Africans in the reserves were encouraged to work on these plots under supervision to improve their farming skills. There were seven such plots in the Molopo and Setlagole reserves.[31]

More popular than the demonstration plots were the bull camps run by the local councils established in 1920 (but only becoming effective in the Bechuanaland reserves about a decade later) and which introduced a better breed of animals into the herds of local stock owners. The Agricultural Department also initiated inoculation programmes and set up dipping tanks. The academic and politician ZK Matthews, who conducted fieldwork among the baRolong, noted that the Mafikeng Council spent £1 000 out of a total budget of £2 275 on agriculture.[32]

The rangers were responsible for ensuring that trees were not felled illegally, and for planting and maintaining new trees.[33] In 1912, regulations were drawn up for the express purpose of conserving the natural resources of the reserves, and the cutting down of trees for wood was meant to be restricted to domestic use – although the legislation probably had little effect as the selling of wood in Vryburg by reserve-based Africans only a few years afterwards appeared to be common. The magistrate in Vryburg reported that:

> There is a vast tract of country exceedingly well wooded between Heuning Vlei and Morokwen. No deforestation has taken place here. At Ganyesa, Morokwen and Takoon [Dithakong], and a part of the Moititon [Boitithong] reserves the natives regularly cut down kameel dorn trees for firewood, and a considerable amount is brought down to Vryburg for sale.[34]

The baRolong farms

The baRolong farms were situated in the southern portion of the Bechuanaland Protectorate and belonged to the Ratshidi. Their possession had been contested by the baNgwaketse and the baRolong and in 1892, following a government inquiry, the baRolong were given the eastern portion, lying between Ramatlabama and Kgoro. A few years later Montshiwa allocated the farms to members of the ruling elite, essentially to ensure their support. The farms were unusual because they were in 'foreign' country (albeit among people closely connected to the baRolong) but divided by a contrived colonial border; and because they were held in individual title, and not communally, as Montshiwa believed that this would safeguard them from white encroachment.[35]

Lotlamoreng Montshiwa was a keen and successful farmer and was credited with introducing 'scientific improvements' and 'cattle improvements' to the Ratshidi early in his career.[36] In the latter part of his life he spent most of his time at his farm Goodhope in the Protectorate, leaving his duties in Mafikeng to his capable headmen Tiego Tawana. Among his many activities Lotlamoreng was running cattle (illegally, in the eyes of the administration) into South Africa.[37] Lotlamoreng was by no means the only progressive baRolong farmer. Others included men such as Modiri Molema, Morara Molema, E Letsaba from Kraaipan, E Masibi from Disaneng, and Tyson Phoi the *kgosi* at Setlagole – a few of those who owned farms privately or in the baRolong farms. Many of these men, including the agricultural demonstrators, sat on the Mafeking Local Council which, as we have seen, devoted much of its energy and expenditure to agriculture and stock-keeping – possibly using the council as a vehicle to advance their own interests.

But for many commoner families it was not all plain sailing in the reserves. Many men and women were forced to work outside in the neighbouring districts, to recoup from poor seasons, or to pay taxes, or raise money for basic foodstuffs. The social anthropologist John Comaroff estimated that, among the Ratshidi, eighty-four per cent of the men and fifty-four per cent of the women had sought work away from Mafikeng for more that ten months at a time.[38] In the more southerly Vryburg reserves, men were frequently reported absent from their homes. An official of the Native Affairs Department, M Meyer, collecting taxes in 1922, observed: 'I found very few men at home, the rest having gone out of the district to look for work.'[39] P-L Breutz, an ethnologist working for the government, and who conducted field work in the mid-1950s, estimated that many of the people were not at home. At Thakwaneng, for example, he found only 1 115 registered taxpayers out of a population of approximately 2 500 to 3 000, and at Klein Chwaing he estimated that only 600 were present out of 1 800.[40]

The number of people seeking work a great distance away is hard to establish, but it appears that much of the work was quite localised. In 1925 the resident magistrate reported that 'with the exception of some Fingos, Basuto, Xhosa and coloured people, all native farm labourers are Bechuana drawn from the reserves',[41] and in the following year the resident magistrate in Vryburg reported that 'the large majority of the able bodied natives go out to work for farmers in this and adjacent districts and at the diamond diggings [around Lichtenburg], they do not go to Johannesburg'.[42] A similar pattern was observed by Jacobs in the Kuruman district.[43] The baTswana disliked underground work on the mines.

'Naboth's vineyard'

If there was a shift in the balance of power between the various political groupings in the Rolong reserves, there was also an internal battle for ultimate authority over the Ratshidi baRolong. There were two 'houses' or centres of power within Ratshidi society: the Tau and the Tawana ruling cliques, headed respectively by the Montshiwa and Molema families and supporters. It was indeed a complex issue because the two families were interrelated. The Molemas had been very influential in the decade or so after the South African War. Both Silas Theleso Molema and his son, the renowned medical doctor, writer and politician Seetshele Modiri Molema, were active in the founding and building of the South African National Native Congress (SANNC, later the ANC).[44]

During these years there was a kind of power vacuum in Ratshidi political life; no particular ruler exercised full control, essentially because the two factions battled for ascendancy. The controversial appointment of Lotlamoreng Montshiwa by the government in 1919 brought the feud to a climax, providing Lotlamoreng with the opportunity to sideline his opponents. In 1933 Lotlamoreng publicly rebuked Sebopiwa Molema (who, up to that point, had been the baRolong national secretary), alleging that funds had been misappropriated from the Ratshidi account and that Sebopia had supported the previous chief, John Bakopoleng Montshiwa, whom Lotlamoreng had displaced. Sebopiwa rejected the charges and responded by charging that 'the chief's object is to lower the Molema status in the tribe after Molema and his elder brother worked harmoniously in the upbringing of the BaRolong tribe … during the lifetime of four successors of the late Montshiwa'.[45] He likened Lotlamoreng's actions to the biblical figure of Naboth who jealously banished his relatives from his vineyard. The Molemas' 'sin' was due to their 'educational qualifications and some acquisition of property'. Sebopia was forced to resign in August 1933. His position and a number of others 'were given to certain favoured headmen of Chief Lotlamoreng'.[46] He went on to accuse Lotlamoreng of spending too much time at his farm Goodhope on the baRolong farms in today's Botswana.

Relations between Dr Modiri Molema and Lotlamoreng were no more cordial. In 1938, when the former attributed an incorrect birth date to the *kgosi* in an article written for the *Mafikeng Mail*, Lotlamoreng took huge exception, and accused him of attempting 'to rekindle the trouble between himself and the Molema section of the tribe'.[47] Despite this, Modiri was lavish in his praise of Lotlamoreng in a brochure he penned for Lotlamoreng's silver jubilee in March 1945. Modiri remained within the inner circle of Ratshidi politics, but did not regain any 'position of eminence'[48] in Ratshidi society. Thus ended any hopes the Tawana might have harboured about placing their preferred candidates in charge of Ratshidi affairs.

The eclipse of the influential Molema family can be interpreted as the end of an intellectual and wider national perspective to Ratshidi-baRolong political life. Sebopia recognised this when he informed the native commissioner in

Mafikeng that 'as a matter of fact the Molema family is hated by their brothers the Montshiwa family by reason of educational qualifications'.[49] Although Lotlamoreng sent a delegation to protest against certain provisions in the Native Adminstration Act of 1927, he generally refrained from close engagement with nationalist politics.[50] Modiri Molema, on the other hand, played an active part in politics, going on to become the treasurer-general of the ANC until, under the Suppression of Communism Act, he was prohibited in 1957 from doing so.

FIGURE 12: Dr Silas Modiri Molema, born 1891 in Mafikeng, died 1965, renowned medical practitioner, author and ANC office bearer.

Source: Molema family

Conclusion

Southern Africa's reserves have been studied by scholars of different disciplines for at least fifty years. They were fundamentally a means by which the more populous African societies could be controlled politically and forced to contribute to the costs of their own governance. From the late 1970s scholars began to see reserves also as labour reservoirs. Men and women could be drawn into South Africa's bourgeoning capitalist system, driven by the mining sector, but their homes would still be in the rural areas. This led to the 'underdevelopment' of local African economies and their 'dependancy on the central economy.[51] This interpretation held sway for a long time.

On closer examination, however, researchers have found that the reserves were more productive than initially thought – they were not just dumping grounds – and that despite all the hardships the inhabitants faced they did survive and some even prospered.[52] The present analysis concurs with such a view. That many baTswana were drawn into wage labour, and at times were forced to eke out a livelihood, did not mean an end to economic life in the reserves – in fact money earned from working outside the reserves could be re-invested back on the land. The chiefs continued to wield control and were looked to for leadership in times of need or stress. Many baTswana in the reserves enthusiastically embraced the new technologies and improvements. Although many did struggle to make a living off the land, and were forced to work for local white farmers, or further afield, the Bechuanaland reserves did not simply collapse. As Jacobs has written of the Kuruman reserves, the local population 'worked creatively and deliberately to mitigate their circumstances and to persevere. Their efforts, and the measure of success they found, are worthy of consideration.'[53]

In addition, political developments and affairs in the Bechuanaland reserves were quite vibrant. Chiefs and local elites competed for local control and engaged with wider political processes and developments outside the narrow boundaries of the territory they occupied. However, by the late 1950s and early 1960s the homeland policy was being put in place, the first stepping stone being the formation of Tswana territorial authorities. Seven different 'authorities' were created in the Molopo/Setlagoli reserves alone, further dividing the communities living there. Not quite knowing what to expect, some of the *dikgosi* accepted the system. Others rejected it. For example, Tshipitota Motseakhumo and his raTlou followers consented to it, whereas *Kgosi* Phoi in Setlagole and the Ratshidi under Tiego Tawana rebuffed it.[54] The system was ultimately to force local chiefs into a much more collaborationist role with the state, and complicated their positions even further. On the whole, too, as was the case of the Hurutshe reserve recounted in the next chapter, it led to further economic decline. From about 1960 it became harder for rural Africans to survive and follow any independent existence in the Bechuanaland reserves.

1   See box on the African Reserves in Southern African history.
2   B.C. Morton, 'A Social and Economic History of a Southern African Reserve: Ngamiland, 1898-1966 (Ph.D thesis, Indiana University 1996, p. 15.
3   N J. Jacobs, *Environment, Power and Injustice: A South African History*, (Cambridge: Cambridge University Press, 2003).
4   See Shillington, *Colonisation of the Southern Tswana*, pp. 194-195.
5   For a more detailed description see P-L Breutz, *Tribes of the Mafeking District*, (Pretoria: Government Printer, 1957) and *Tribes of the Vryburg District* (Pretoria: Government Printer, 1959).
6   Cape Archives Depot, (CAD), 1/Mfk, 7/1/12, Native Commissioner, Mafikeng to Lotlamoreng Montshiwa, 9 September 1925; 1/Mfk, 7/1/5, N 4/6/7, 'Circular to all Magistrates re. Stock losses from Dept. of Agriculture', 10 April 1934.
7   CAD, 1/VBG, vol. 17/29, r 2/1/10, Rev. Peasley to NC, Vryburg, 16/3/1933.
8   CAD, 1/VBG, vol. 17/31, r 2/4/1, Visit of the Native Commission, 19/4/1923.
9   CAD 1/VBG, 17/25, 618/14/33, N.C. Vryburg to R.M. Vryburg, 22/07/ 1925.
10 CAD 1/VBG, 11/13. 2/3/4, R.M. Vryburg to SNA 11/09/1225.
11 Shillington, *Colonisation of the Southern Tswana*, p.138.
12 For accounts of the Rietfontein dispute and its resolution see M.D. Ramaroka, 'The Early Internal Politics of the BaRolong in the District of Mafikeng: Intra-Batswana Ethnicity and Political Culture from 1852-1920.' (M.A. Thesis, North-West University, 2003), Molema, *Montshiwa*, p. 110-116 and editions of the *Kimberley Advertiser*, April and May 1918, especially for 25 May 1918.
13 Ramaroka, 'Early Internal Politics of the BaRolong', 138.
14 NTS. vol 108.
15 See S.M. Molema, *Montshiwa*, pp. 70-71.
16 Molema-Plaatje Papers, A9, C2, Disputes, Circuit Court, Mafikeng, in matter between John Masibi, (Appellant) and Pertus Masibi (Respondent).
17 CAD, 1/Mfk, 7/1/14, N1/1/5, Report to Resident Commissioner, Bechuanaland Protectorate, from Assistant Commissioner, Mafikeng, 6 Feb 1932.
18   CAD, 1/Mfk vol 12, n1/1/3 (1) part 2, Lotlamoreng Montshiwa to Chief Native Commissioner, Northern Areas, 20 April 1939.
19 CAD, 1/Mkf 7/1/14, N1/1/5/2 part 2, NC Mafikeng , G. van H. Tulleken, to Chief NC Pretoria, 9 Nov 1946.
20 CAD, 1/Mfk, 7/1/72, N2/11/2, NC Mafikeng to Chief NC, Potchefstroom, 2 October 1951, 'Rehabilitation of Native Areas'.
21 CAD , 1/Mfk, 7/7/1/1/11 N2 /3/5, Contained in NC Mafikeng (Eaton) to Chief NC Western Areas, 2 September 1950.
22 CAD, 1/Mfk, 7/1/72, N2/2/3/2, J. Melman to NC Mafikeng, 12 August 1956.
23 CAD 1/Mfk, 7/1/1/11 NC Mafikeng to SNA, 15 September 1956.
24 CAD, 1/Mfk, 7/1/72, N2/2/3/2, NC Mafikeng to Chief NC Potchefstroom, 4 June 1952.
25 Jacobs, *Environment, Power and Injustice*, 123.
26 See CAD, 1/MFK, vol. 7/1/97, r. N8/21/4, NC Mafikeng to Chief NC, Potchefstroom, 8/7/1949, 'Chart of Native Livestock'; 1/MFK, vol. 7/1/97, r. N8/212/4, Temporary Agricultural Officer, to Director of Native Agriculture, 30/10/1943.
27 CAD, 1/MFK, vol. 7/1/102, r. N4/3/1, 'Census of Bantu Agriculture, Mafeking, 1959-1960, compiled by Agricultural Officer, n.d.

28   CAD, 1/VBG, vol. 17/6, r.6/17/2, 'Estimates, Natives and Native Stock, Vryburg' n.d.

29   CAD, 1/MFK, vol. 7/1/86, r. N5/1/3, 'Boreholes Operations Progress, Mafeking District, 1942.

30   CAD 1/VBG, vol. 3/45/2. r. 5/12, NC Vryburg to SNA, 12 March 1936.

31   See J. Drummond, 'Changing Patterns of Land Use and Agricultural Production in Dinokana Village, Bophuthatswana' (M.A. University of the Witwatersrand, 1992, pp. 82-86.

32   Z.K. Matthews, 'A Short History of the Tshidi-BaRolong', *Fort Hare Papers*, (June 1945), p. 25.

33   CAD, 1/MFK, vol. 7/1/97, r. N8/21/4, Temporary Agricultural Officer to Director of Native Agriculture, 30/10/1943.

34   CAD, 1/VBG, vol. 17/23 r. 307/14/19, R.M. Vryburg to Conservator of Forests, 3/11/1915, 'Deforestation in Bechuanaland', 1914-1916.

35   See Molema, *Montshiwa, Barolong Chief and Patriot*, p. 193.

36   Molema-Plaatje Papers, A 2 BaRolong Research, '*The First Twenty-Five Years of the Chieftainship of Lotlamoreng Montshiwa* by S. M. Molema, contained in Notes on the Chieftainship of Lotlamoreng by Z.K. Matthews, William Cullen Archive of Historical and Literary Papers, University of the Witwatersrand.

37   CAD, 1/Mfk, vol. 7/1/12 r. N1/1/3 (part 1), Proceedings at an Enquiry Under Section 96 of Act 31 of 1917, Held at Mafeking, 5 /12/1932.

38   J. Comaroff, 'Competition for Office and Political Processes among the Barolong boo Ratshidi of the South African-Botswana Borderland, (Ph.D thesis, University of London, 1973) p. 54.

39   CAD 1/VBG, vol. 17/31, r. 2/4/1, M. Meyer (Clerk) to RM Vryburg, 22/5/1922.

40   Breutz, *Tribes of Vryburg District*, 154-154.

41   CAD 1/VBG, vol. 17/44, r. 8/4/2, RM Vryburg to Secretary for Agriculture, 9 December, 1925, 'Native Labour on Farms'.

42   CAD 1/VBG, vol. 17/32, r. 2/5/4, RM Vryburg to SNA, 24 September 1924.

43   Jacobs, *Environment, Power and Injustice*, 134.

44   See A. Odendaal, The Founders: The Origin of the ANC and the Struggle for Democracy in South Africa (Johannseburg: Jacana, 2012). p. 177. Odendaal credits the baRolong with playing 'an important part in African organizational politics in South Africa'.

45   CAD, 1/Mfk, 7/1/12, N1/2/5. S.J. Molema to NC Mafikeng, 7 June 1933.

46   CAD, 1/Mfk, 7/1/12, N1/2/5. S.J. Molema to NC Mafikeng, 16 August 1933.

47   CAD, 1/Mfk, 7/1/16, N1/1/3 (1) part 2, NC Mafikeng to Chief NC, Northern Areas, 12 April 1938. This may not have been an innocuous mistake by Dr Molema, as questions about Lotlamoreng's paternity and birth after his father's death were central to attempts to delegitimise his right to the chieftainship.

48   Comaroff, 'Competition for Office and Political Processes among the BaRolong Boo Ratshidi of the South African-Botswana Borderland', (Ph. D thesis, University of London, 1973), p. 391.

49   CAD, 1/Mfk, 7/1/12, N1/2/5, S.J. Molema to NC Mafikeng, 7 June 1933.

50   See S.M. Molema, 'The First Twenty-Five Years of Chieftainship', Brochure to mark Silver Jubilee Celebrations. Z.K. Matthews Collected Papers, A2 Barolong Research, University of South Africa Archives.

51   The 'underdevelopment' paradigm, in the southern African context, derived principally from the writings of G. Arrighi, *The Political Economy of Rhodesia* (The Hague:

Mouton, 1967), and H. Wolpe, 'Capitalism and Cheap Labour Power in South Africa: from Segregation to Apartheid, *Economy and Society*, 1, (1972), 425-456.

52 See for example, I. Phimister, 'Rethinking the Reserves: Southern Rhodesia's Land Husbandry Act Reviewed', *Journal of Southern African Studies*, 19, 2, (1993), pp. 225-239, and B. Morton, 'A Social and Economic History of Ngamiland'.

53 *Jacobs, Environment*, Power and Injustice, p. 207.

54 CAD, 1/Mfk, 7/1/111, N11/1/1, NC Mafikeng to Chief NC, Potchefstroom, 6 January 1955. The Bantu Authorities Act was passed in 1951.

CHAPTER 5

Rural resistance: The baHurutshe revolt of 1957-58

The 1957 baHurutshe revolt

In 1957 and 1958 a majority of the residents in Moiloa's Reserve, both men and women, mounted sustained resistance to the issuing of passes for women in the reserve. This has rightly been seen as a telling event in the history of the country's rural resistance to white rule, and one that has an iconic place in the grand narrative of the ANC's resistance history. The generally accepted view is that it was the attempt to force African women to carry passes that triggered the resistance, and the event is rightly a significant component of the history of women's resistance in South Africa – which would imply that the grievances experienced by the reserve's residents were relatively short term from the time the carrying of passes became compulsory in 1954. In this chapter, however, we argue that the trouble had been brewing for many years and the revolt can only be fully understood by examining developments in Moiloa's Reserve from several decades before the actual revolt broke out. The baHurutshe revolt was not primarily about pass resistance, but about retaining access to vital resources in the baHurutshe reserve. In this sense it bears a strong resemblance to the perhaps better known revolt in Sekhukhuneland which broke out shortly afterwards. Similarly, the full effect of this incident can only be comprehended by an analysis of the subsequent course of events over the ensuing decades.

The best account of the events that unfolded in Zeerust was recorded by the Reverend Charles Hooper, the Anglican priest in Zeerust.[1] He personally intervened to assist the baHurutshe, for which he was finally deported from South Africa. Other commentators have examined specific aspects of the event, focusing for example on the role of the ANC, and the extent of political consciousness exhibited by the women in the reserve.[2]

In March 1957, Abram Ramotshere Pogiso Moiloa, the ruling *kgosi* in Dinokana, was instructed by Carl Richter, the native commissioner in Zeerust, to tell the women of Dinokana to present themselves at the *kgotla* to collect their reference books.[3] By this date, the issuing of passes had been strongly resisted by both urban and rural women throughout South Africa. In 1956 alone, 'the year when "reference books" first began to be issued, some 50 000 women demonstrated against the pass laws on thirty-eight different occasions in thirty different places'.[4] Abram reacted first by summoning his councillors and then by directing a woman's representative from each ward to meet with him for discussions. According to informants, he decided to reject the reference books in the course of this meeting.[5] Abram simply ignored the order from Richter, who seized the opportunity to depose him as *kgosi*.

Abram did not move out of the reserve as Richter had allegedly ordered him to do, but went into hiding in Dinokana. A pass-issuing unit then set up business in the shop of a white trader, but according to one of the participants in the resistance only eighty out of the approximately 8 000 women in the Dinokana area took out passes.[6] On 6 April a delegation of women came from Johannesburg and encouraged others not to accept the passes, and on 13 April a group of men from the Witwatersrand, who had formed a support group known as the Bahurutshe Association, chartered two buses and returned to the reserve to investigate the situation. This precipitated a more violent phase in the resistance in which the houses of a number of pro-government chiefs and their supporters were burned down. Four men, including Michael Moiloa, who is mentioned below, were seen as the key collaborators and were blamed for the state's repressive antics. A police contingent had to be sent to the reserve to prevent them from being 'thrown into a deep hole (named Mamakgodi) in a nearby hill'.[7] Passes were collected up and publicly burned. In nearby Gopane, *Kgosi* Albert Gopane was forced to flee to Botswana after townsfolk condemned him for accepting the pass system.

From its side, the state acted quickly to quell the disturbances. The returning buses (of the Bahurutshe Association) were stopped and a number of people arrested. Postal and transport services in the reserve were withdrawn, and resisting women were denied access to medical treatment. These intimidating tactics were soon backed up by extreme force. By mid-April a mobile column under the command of a notoriously brutal police sergeant by the name of Van Rooyen combed the reserve seeking the women who had refused or burned passes. The round-up continued for several months, with the state hoping to enforce

FIGURE 13: The legendary Nye Nye tree in the Dinokana *kgotla*, where meetings were held to plan a response to the crisis in Lehurutshe in 1957.

Source: The authors

acceptance of the passes through the courts. It was not particularly success-ful; by October 474 people had been arrested, but only thirty-nine convictions obtained. As the women of Gopane had shown by their *en masse* refusal to take up passes, this tactic was quite effective in an area where there were few police-men and limited prison accommodation. The failure to obtain convictions was also – and mainly – because the baHurutshe enlisted that services of Shulamith Muller (one of the instructing attorneys in the 1956 Treason Trial) and Advocate George Bizos, later a stalwart of the legal battle against the apartheid policies of the National Party government. The state indicted Bizos's clients for the crime of *crimen laesae majestatis*, 'usurping the function of the state by establishing and conducting a court',[8] specifically that in which the collaborators had been found guilty of causing the unrest.

Ramotshere was still in hiding, and attempts by Richter to appoint another *kgosi* had come to nothing, owing to the determined opposition of the residents of Dinokana. In October the Native Affairs Department then appointed a com-mission, the only one until then to be appointed to investigate a rural trouble spot in South Africa, and the government banned gatherings of more than ten people in the reserve. The road from Dinokana to Zeerust was sealed off on 15 October and people attending the hearings were assaulted and harassed. Harvard aircraft swept the reserve in a blatant act of intimidation. Apart from the extraordinary brevity between the gazetting of the commission and its sitting, it was unusual in other ways: there was no agenda, there was nobody to lead evidence, and Advocate Bizos was not allowed to cross-examine witnesses.

Up to this point the resistance had been confined to the villages of Dinokana, but after November it spread to Braklaagte, (Lekubu) Witkleigat (Moshana), Motswedi and Leeuwfontein, encompassing practically the entire reserve. Resistance was strongest where the local *dikgosi*, such as Lucas Mangope in Motswedi and Edward Lentswe in Witkleigat, had supported the state. The motives of such *dikgosi* may have been to avoid a confrontation with the state but it led to a spiral of violence and counter resistance. The mobile column moved into the new trouble spots in late 1957 to lend assistance to the compliant *dikgosi* and their local bodyguards.

The real backlash came on Christmas Day. The mobile column was off-duty and many men of the Bahurutshe Association were home for the holiday period. Lencoe's house and car were burned, his wife was assaulted, and he was forced to flee to Botswana on horseback. Thirty-six houses belonging to alleged government supporters were burned in Witkleigat and fifteen in Leeuwfontein. Lentswe's right-hand man was killed during these disturbances. In Motswedi, *Kgosi* Mangope suffered an attack on his house and person, and his household was forced to open fire on a mob of enraged villagers.

The police responded quickly. Witkleigat was surrounded and the search for the perpetrators was launched. At Motswedi, a temporary court was erected and placed under police guard. Subsequently, Mangope (still suffering the effects of his attack) appeared in a dressing-gown to impose fines for pass-burning and arson. In Gopane, the newspaper *New Age* reported that: 'Public trials and confession are the order of the day. Men are made to apologise for being [African National] Congress-ites, and then enrolled as the chief's forces.'[9] Matters reached a tragic climax on 25 January 1958 when investigating policemen in Gopane were mobbed and four innocent bystanders were shot and killed in the ensuing panic.

The shootings shocked the inhabitants, and resistance came to an end. Mass arrests were made and about 200 people were charged with murder. By September, five had been convicted of assault and fifty-eight of public violence, and the remainder were acquitted. According to Bizos, most of the women charged with pass burning were acquitted 'because the prosecutor was unqualified and inexperienced, because the police work was sloppy, and because of the lack of sophistication of the elderly magistrate'.[10] Fearful of what the future might hold, between December 1958 and February 1959 many fled their homes, up to a thousand crossing over to Botswana (then still the Bechuanaland Protectorate), where they were fed and clothed for eight months by the office of the British High Commission. Many never returned to South Africa. *Kgosi* Abram Moiloa was reported to have crossed into Botswana in late January 1958 and to have sought protection for himself and a number of his followers from the bamaNgwato paramount Tshekedi Kgama at Pilikwe. For several weeks, hundreds of Mohurutshe men and women sought sanctuary with the Hoopers at the Anglican church in Zeerust. The shootings and Moiloa's flight to Botswana undoubtedly had a demoralising

FIGURE 14: St John's Anglican Church, Zeerust, where many of the baHurutshe found refuge with the Reverend Charles Hooper and his wife Sheila.

Source: The authors

effect on the entire reserve but they were not the only reasons for the collapse of the revolt. The leaders of the revolt were rounded up and deported; the ANC was banned in the reserve in March 1956; and in Johannesburg the police arrested the so-called 'trouble makers' of the Bahurutshe Association, and threatened them with detention. In addition, many people had suffered financially over the past year and wanted to resume the planting and harvesting of their crops, and to tend their livestock.

Information about events after April 1958 is scant. Charles Hooper and his wife Sheila, who throughout the crisis had been supportive of the baHurutshe, were deported from Zeerust. No reporters were allowed into the reserve. However, in August 1959 the Bahurutshe Regional Authority, the first of its kind in the Transvaal, was proclaimed on a trust farm about forty kilometres north of Zeerust. The system of bantu authorities (mentioned in the previous chapter) was being fully implemented in the reserve. An uncle of Abram Moiloa's, Israel, was installed as *kgosi* in Dinokana, but the ceremony was presided over by Lucas Mangope who on this occasion delighted his critics by 'imploring the minister of bantu administration to "lead us and we shall try to crawl"'.[11] This was to be a significant milestone in the career of Mangope as it linked him irrevocably with developing bantustan policy. It was an unforeseen development, for Mangope had been a founder member of the South African National Native Congress (SANNC), forerunner of the ANC. His son, also Lucas, went on to become the first and only president of the ill-fated Bophuthatswana homeland. George Bizos, on the other hand, 'was invited to become an honorary member of the [baHurutshe] tribe'.[12]

Commentators have picked up on the role of the Bahurutshe Association as an example of the way in which the migrant system in South Africa linked the urban and rural African population. The part played by the Bahurutshe Association in the revolt bears striking similarities to that of migrants in Sekhukhuneland, who 'forged networks linking urban and rural areas … [and] kept their home communities abreast of developments'.[13] Walter Sisulu, elder statesman of the ANC, has described the Sekhukhuneland resistance as 'the first well organised [rural] movement in the history of the ANC'.[14] However the Zeerust revolt preceded it in the then northern Transvaal, and there is evidence that the Sekhukhuneland migrants drew on the experiences of the Bahurutshe Association.

Hooper and some other observers have played down the extent of the ANC's role. This may have been to protect the baHurutshe against the state's frequent assertion that they were mere docile dupes in the hands of ANC instigators. Others have used the Zeerust revolt to suggest that the ANC was not fully aware of, or responsive to, the needs of rural people. The evidence suggests otherwise. In an interview in 1983 Kenneth Mosenyi explained how he organised a branch of the ANC in Dinokana, and played a leading role in the revolt.[15] Documentary evidence now made available from the perspective of the state, and oral recollections now provided without fear of recrimination, show the extent of ANC presence in the reserve. In addition to Mosenyi, Nimrod Moagi, who lived near Dinokana, was a long-standing ANC member (both men were probably full-time paid ANC officials). Mosenyi had been a trade unionist and had personal contact with a number of resistance luminaries such as JB Marks and Dr Y Dadoo, and became an executive member of the Council of Non-European Trade Unions (Nactu).[16]

In addition to these men, Boas Moiloa, the chief's uncle, was said to have been 'in close cooperation with the African National Congress even long before Abram was deposed'. Simon Molifi 'a known agitator of unknown status was apparently

FIGURE 14: *Kgosi* Ramotshere Moiloa, after his return from exile in the late 1970s.
Source: BaHurutshe Tribal Office

tasked with assuming responsibility for 'propogating [sic] communistic ideas should Kenneth Mosenye be prevented to [sic] continue with his undermining activities in Dinokana', and another leading Hurutshe councillor, David Moiloa, was identified as '*n genoemde kummunis*.[17]

It is significant that apparently Mosenyi introduced Moiloa to the ANC. The security police records reveal that it was 'under his [Mosenyi's] influence the old Kaptyn Moiloa travelled with him to attend meetings with the ANC', clandestinely in Johannesburg. The culmination of Moiloa's politicisation and identification with the ANC came when he sent a letter of support to the Congress of the People convention in Kliptown in 1954 (the letter was intercepted by the police). Interviewed in 1983, two of the leading figures among the women of the Hurutshe reserve, Joanna Pule and Paulina Keebine, recalled that a number of ANC Women's League office holders, in particular Lillian Ngoyi, were their acknowledged 'leaders'.[18]

The women of Moiloa's Reserve were familiar with the growing national anti-pass movement. In Johannesburg, a number of Hurutshe women were members of the Women's League, and urged women in the reserve to refuse the passes and to adopt the boycott tactics as in the recent Rand bus boycotts. Once the resistance gained momentum, the ANC acted fairly speedily. Shulamith Muller spent many days in Zeerust after legal assistance had been sought, and Nelson Mandela himself visited prisoners awaiting trial in the Zeerust jail.[19] Ramotshere Abram Moiloa has been elevated to the status of a 'struggle hero' by the post-apartheid ANC government (a municipality in the Zeerust district is named after him). Not only did he expose himself to the dangers inherent in resisting the all-powerful white regime, before and during the revolt itself, but when he went into exile in Bechuanaland he made contact with the ANC after it had been banned in South Africa and directly recruited a number of young men from Dinokana who served in the celebrated Luthuli detachment of the ANC's military wing.[20]

The longer view

There were other reasons, too, for the prominent role of women in the revolt. In order to grasp the full import of and deep-rooted anger behind the revolt, it is necessary to backtrack to as far back as the beginning of the twentieth century.

It was recorded in the chapter on *Kgosi* Moiloa that the reserve, had been granted to the chiefdom in 1848 by then President Pretorious. Its size had been increased by the purchase of several farms on its borders. Although the state had interfered increasingly in routine matters affecting the lives of the residents on Moiloa's reserve, by 1915 the productive base of the reserve was still very much intact.[21] However, there had been a shift away from household production under the control of women into the hands of what the authorities called 'progressive farmers', all males (very similar to the situation in the Bechuanaland reserves). Increasingly, in the 1920s and 1930s, these 'progressives' and the traditional authorities came into conflict. The interests of the progressives hinged on two key figures, Michael and Joseph Moiloa. Michael had attended the local Lutheran mission school up to Standard 8, was a warden in the church and, by the 1920s, had become a prosperous citrus farmer on his lands around Dinokana. From the early 1920s Michael and his supporters attempted to control local politics in the reserve. The Zeerust Native Affairs Department was quick to notice this and threw its weight behind the acting *kgosi*, Alfred Moiloa. The local commissioner described Michael as a man '… of whom I have every reason to hold a very mean opinion, and one who surreptitiously makes trouble. As he is the leading man among the mission station, he has … influence.'[22] In the late 1920s, the progressives began to push for the formation of a Moiloa Reserve local council under legislation passed in 1920, in order to gain control of the budget which the council controlled (and was raised from local taxes). The council, established in 1931, could make proposals to the authorities regarding matters of local interest such as health services, road maintenance, agricultural and pastoral issues (irrigation measures, control of stock diseases, sites of grazing camps) and became a 'useful body through which the progressives could advance their own interests'.[23] Whereas the chiefly representatives wanted to use the council to obtain traditional dues (such as taxes on migrants) the progressives wanted to improve agriculture and encourage entrepreneurial activities in the reserve.

In the 1930s, however, the state's interest in the reserve began to shift towards the 'progressives'. In 1929 the Native Affairs Department appointed its first director of native agriculture, RW Thornton, as part of 'a general strategy to arrest the collapse of the reserves'.[24] The changing nature of the Native Affairs Department from the 1930s has attracted attention from other writers interested in the reserves and other black rural districts. The historian Peter Delius has observed that there was an 'infusion of agriculturalists', well educated and professional, and mostly English-speaking, who were appointed to work in them.[25] Consequently, in Moiloa's Reserve the new sub-department stressed modern animal husbandry, soil conservation, and efficient agricultural techniques. These measures were part of government policies to make the reserves self-sustaining, so that Africans were not a burden on the state and to promote a class of African peasant farmers in the reserves – thus strengthening the position of Michael and the progressives in Moiloa's Reserve and making it easier for them to gain the upper hand in the local council. It was a situation that satisfied the government. As the local native commissioner was to note in 1936, 'with the establishment of the Moiloa Reserve Local Council the whole reserve bids in a fair way to become "a model native area"'.[26] Michael became a 'trusted confidante' of the Native Affairs Department and was a strong supporter of the betterment plans for the reserve.

In 1933 the new young *kgosi*, Ramotshere Abram Moiloa, assumed control of the baHurutshe at Dinokana. Having observed the gradual erosion of the chief's powers under Alfred's regency, many now looked to Abram to restore the traditional order. Michael Moiloa was excluded from the *khuduthamaga* and Ramotshere began to tighten up on the customary dues, which he could levy through various means, mainly on migrants leaving the reserve. He also led the resistance to the betterment policy (in particular, a plan to fence the reserve). Eventually his activities annoyed the native commissioner who reported, in 1939, the 'impossible conduct of Chief AG Moiloa', and requested the Native Affairs Department to discipline him. 'As the chief's attitude makes proper administrative control impossible, I … suggest nothing short of summary dismissal,' he urged.[27] At this time however the 'tide of state policy was flowing towards a more substantial role for chiefs' and the *status quo* was maintained[28] even though there was clearly tension between the progressives and the chiefly authorities, the balance of power lying with the progressives, who now had greater support from the government. As long as economic opportunities and production continued to exist in the reserve, however, these tensions did not reach a point of open conflict.

But almost immediately after the National Party came to power in 1948 there was a noticeable decline in the reserve's economy. Indeed, no less a figure than HF Verwoerd, the key architect of apartheid himself, assumed control of the Native Affairs Department and Dr WW Eiselen took over as the secretary for native affairs. The new order was more bureaucratic and its officials were enthusiastic proponents of separate development. Eiselen cut down on the wartime system of advancing credit to African reserve-based farmers. In 1955, the Tomlinson Commission was established to examine overcrowding in the reserves which had resulted in uneconomic landholdings, low levels of production and environmental degradation. The commission recommended an extension of land holdings to a class of progressive farmers.

It was thus the sudden threat of the loss of land, coupled with the simmer-

FIGURE 15: Ramotshere Moiloa at the date of his appointment in 1933.
Source: BaHurutshe tribal Office, Dinokana

ing tensions between the progressives and traditional leadership, that created the conditions for the 1957 revolt.[29] The gist of this argument is that there is evidence that, from the 1940s, women in particular were increasingly displaced or marginalised as cultivators. Michael Moiloa, for one, supported this, stating that 'women should not be given a chance [to cultivate demonstration plots] as they are always occupied by domestic affairs at their homes, but … the young men should be given the opportunity'.[30] On a wider political level, the South African reserves, as one writer has put it, 'enjoyed pride of place in the [apartheid government's] vision of the future' and the state played an 'increasingly interventionist role in the countryside' from the early 1950s. [31] From the government's point of view, chiefs such as Abram Moiloa, who had blocked resistance to betterment, for example, were among those who should be replaced by more compliant leaders.

Thus by 1955 *Kgosi* Moiloa's position was threatened by both the Native Affairs Department and the 'progressive' group in the reserve. He accused them of undermining his authority by inciting people to appeal against his judgements and cultivating land without his permission, accusations which the Native Affairs Department rejected out of hand. Abram was also outspoken against the Bantu Authorities Act and the Bantu Education Act, cornerstones of the National Party's apartheid policy. 'Who the hell is Verwoerd?' he is claimed to have asked. 'He is just a minister. I am not afraid of him.'[32] In addition, he opposed the removal of the 'black spots' of the villages of Braklaagte and Leeufontein, the two baHurutshe settlements just outside the reserve. He was described as a 'thorn in the flesh of the Department [of Native Affairs]' and placed under investigation 'with a view to requesting the department to remove him from the chieftainship and banish him from Zeerust'.[33] A subsequent inquiry, conducted by the native commissioner for Pilanesberg, found Abram Moiloa 'guilty' of eighteen charges of misconduct prior to the Hurutshe revolt itself.

In May 1956 the newly-arrived native commissioner in Zeerust, one Carl Richter, accused Abram of 'interfering with the police, the messenger of the court, the postal officials, the churches and the local farmers' and stated quite categorically that he wished to remove Abram: '… so that we can declare Moiloa's Reserve a betterment area, but with Abram in the reserve he will mobilise people against it … it is clear too that he is under the influence of Mosinyi and Nimrod Moagi instead of experienced men like Michael Moiloa'.[34]

Thus the revolt should be seen in 'the context of a longstanding struggle with the progressive farmers of the reserve, who by the mid-1950s were in a close alliance with the state',[35] whereas Abram rallied support among the women subsistence producers, particularly in Dinokana, who saw him as the defender of their rural livelihoods. As Bundy and Beinart have argued for the Transkei, the institution of chieftaincy 'served as a powerful ideological force when rural resources or political structures came under threat'.[36] It is no surprise that the women saw the issuing of the dreaded reference books, which controlled their movements to and from the urban areas, as the final straw in a long campaign to deprive them of a secure means of support.

The effects of the baHurutshe revolt

In the longer term, the revolt hastened political transformation in the reserve. The leadership crisis (caused by Abram's flight and his uncle Boas's refusal to take over) led to the appointment of Michael Moiloa himself in September 1958. His appointment, together with Mangope's rise to prominence, lucidly illustrate Delius's observation that Bantu authorities 'la[id] the foundations of a political and economic arena which would absorb the energies and aspirations of the African middle class'.[37]

However, Michael was an unpopular choice and he eventually lost control, asking to be relieved of his duties in 1960. His departure occasioned another crisis. The government, having intervened so decisively in baHurutshe affairs, was determined not to restore to power any of the former leaders associated with the old order. Eventually, on the recommendation of government ethnologists, Israel Moiloa from Braklaagte, a member of the ruling family, filled the vacant chieftaincy in Dinokana. As noted above, other traditional leaders who had sided with the state (such as the Mangope family) and whom the authorities regarded as supportive of state policy (especially the betterment scheme) were elevated to positions of authority. However, both Ramotshere Moiloa and David Moiloa, on their return from exile in different parts of southern Africa, were to confront the authorities over their attempts to install a pro-government chieftaincy in the neighbouring village of Leeuwfontein in the late 1970s, an indication perhaps of the support they still enjoyed among the baHurutshe.

The installation of a more compliant leadership also supported the state's economic plans for the reserve, in particular betterment and the fencing of the reserve. The disturbances temporarily curtailed agriculture in the reserve. Many people went into hiding, or fled across the border to Botswana, never to return; whether their lands were reallocated to others is not clear. Though data is hard to obtain, the dislocation caused by the revolt probably forced more men into a pattern of labour migrancy and, in turn, women became more reliant on remittances from men working in occupations outside the reserve. Apartheid policies such as influx control and forced removals added substantially to land hunger during the 1960s and 1970s. The settlement of Lehurutshe (Welbedacht) was created in 1969, when labour tenants and 'squatters' were moved there from white farms in the western Transvaal. These removals, 'added over thirteen thousand people to the reserve's population, contributing to overpopulation, the erosion of pastures, and a fall in crop yields'.[38] In addition, water, formerly a plentiful resource in Dinokana, was redirected to the settlement area at Welbedacht, which affected the flow of water from the Dinokana spring and restricted irrigated cultivation in the vicinity. Gradually there was a noticeable contraction in the area of land under cultivation in the south of the reserve – from 470 hectares in 1957 to 206 in 1984. The resilient economy of Moiloa's reserve had finally collapsed.[39]

Ramotshere Abram Moiloa, wishing to die at home in South Africa, applied to return to Dinokana in 1971, but the government refused his request. A second application in 1975 was successful and he was allowed to return home. By this time he was an old, frail and sickly man. As research by Lissoni has revealed, he was actually reappointed as *kgosi* at the request of Mangope who, it seems, was 'trying to give the institution of chieftaincy a new lease of life'[40] or maybe trying to buy some credibility for the fledgling Bophuthatswana state. He died in 1982.[41] From the 1960s the Lehurutshe reserve, bordering on Botswana, became well known as an exit route for people going into exile, and for supporting insurgents entering into South Africa. Interest in the ANC that had declined to some extent during the 1930s and 40s began to revive among many of the occupants of the western bushveld. These were political legacies of the Zeerust revolt.

ENDNOTES

1   See C. Hooper, *Brief Authority*, (London 1960).
2   See A. Manson, 'The Hurutshe Resistance in the Zeerust District of the Western Transvaal, 1954-1959', *Africa Perspective*, no. 22 (1983); J. Yawitch, 'The Zeerust Revolt (unpublished seminar paper, University of the Witwatersand, 1982).
3   This is drawn from most of the available sources, in particular Hooper's *Brief Authority*, Fairburn's article in *Africa South*, South African Institute of Race Relations, *Survey of Race Relations, 1957-58*, and various newspapers, (*New Age, Fighting Talk and Drum*).
4   C. Walker, 'The Federation of South African Women 1954-1962', Collected Conference Papers on History of Opposition in Southern Africa, p.183.
5   See Interview with Mrs. M. Keebine, South African Institute of Race Relations Oral Archive, Interview by A. Manson, 4 December 1982.
6   *New Age*, 28 November 1957, from evidence given to the Commission of Enquiry.
7   Cited in G. Bizos, *Odyssey to Freedom*, (Johannesburg: Random House, 2007), p.106.
8   Bizos, *Odyssey*, p.112.
9   *New Age*, 13 February 1958.
10 Bizos, *Odyssey*, p.113.
11 Cited in T. Lodge, *Black Politics in South Africa from 1945*, (Johannesburg: Ravan,

1983*)*, from the *Cape Argus*, 9 August 1959.

12 See, Bizos, *Odyssey*, p.117.

13 P. Delius, '*A Lion Among the Cattle': Reconstruction and Resistance in the Northern Transvaal*, (Johannesburg: Ravan, 1996), pp. 48, 83 and 113.

14 Cited in Delius, '*A Lion Among the Cattle*', p. 103.

15 Interview with Kenneth Mosenyi, by Andrew Manson, 18 January 1983. SAIRR Oral Archive, William Cullen Library, University of Witwatersrand.

16 See NTS vol. 326, r 40/55, Memorandum: Kenneth Benjamin Mosenyi.

17 NTS vol. 326, r 40/55, Senior Investigator, South African Police to Chief NC, Potchefstroom.

18 SAIRR Oral History Archive, Interview with Johanna Pule, by Andrew Manson and Tebogo Kubine, 3 March 1983.

19 These facts are set out in Manson, 'The Hurutshe Resistance' pp. 72-75.

20 Personal communication with Arianna Lissoni, Department of History, History Workshop, University of the Witwatersrand, 7 April 2011. Lissoni has interviewed a number of these men.

21 See Manson, 'The Hurutshe in the Marico District', pp. 188-226.

22 Records of the Native Affairs Department, (NTS), Transvaal Archives, vol. 37/360, NC Zeerust to SNA, 17/11/1930.

23 J. Drummond and A. Manson, "The Rise and Demise of African Agricultural Production in Dinokana Village, Bophuthatswana', in *Canadian Journal of African Studies*, vol. 27, no 3, 1993, and J.H. Drummond, 'Changing Patterns of Land Use and Agricultural Production in Dinokana Village, Bophuthatswana', M.A Thesis, University of the Witwatersrand, 1992, pp. 73-84.

24 S. Dubow, "'Holding a Just Balance between Black and White", The Native Affairs Department in South Africa, c. 1920-1933', *Journal of Southern Africa Studies*, 12, 2, (1986), p. 230.

25 P. Delius, *A Lion Among the Cattle*, p. 48.

26 NTS vol. 8537, NC Zeerust to Secretary for Lands, 8 July 1932.

27 NTS vol. 325 r140/162, Assistant NC to Chief NC, 25/10/1939.

28 W. Beinart, 'Agrarian Historiography and Agrarian Reconstruction', in J. Lonsdale (ed.), *South Africa in Question* (Cambridge: Cambridge University Press, James Currey and Heinemann, 1988), p. 137.

29 This is the gist of the argument put forward by Drummond and Manson, 'The Rise and Demise of African Agricultural Production'.

30 NTS vol. 325 r140/162, Assistant NC to Chief NC, 25/10/1939.

31 P. Delius, *A Lion Among the Cattle*, p. 76. See also T. Lodge, *Black Politics in South Africa Since 1945*, (Johannesburg: Ravan, 1985), p. 261.

32 See J. Fairbairn. 'Zeerust: A Profile of Resistance', *Africa South*, (April/June1956). P. 17.

33 NTS vol. 326, r140/55 Chief NC to NC Zeerust, 18/02/51.

34 NTS vol. 432, Native Commissioner, Zeerust to Magistrate Zeerust, 19/05/1956.

35 Drummond and Manson, 'The Rise and Demise of Agriculture', p. 472.

36 W. Beinart and C. Bundy, 'State Intervention and Rural Resistance: The Transkei, 1900-1965', in M. Klein (ed.), *Peasants in Africa, Historical Contemporary Perspectives*, (Beverly Hills: Sage, 1980).

37 Delius, *A Lion Among the Cattle*, p. 78.

38 Drummond and Manson, 'The Rise and Demise of Agriculture' p. 475.

39 See J.H. Drumond, 'Changing Patterns of Land Use and Agricultural Production

in Dinokana Village, Bophuthatswana', (M.A Thesis, University of the Witwatersrand, 1992), pp. 140-155.

40 A. Lissoni, 'The Bahurutshe chieftaincy and Bophuthatswana c.1975-1989', unpublished paper, University of the Witwatersrand, 2011, p. 9.
41 Personal communication with A. Lissoni, 7 April 2011.

CHAPTER 6

'Blunting the prickly pear': Bophuthatswana and its consequences 1977-1994

Beware that Bophuthatswana is like a prickly pear. It is very tasty but it is also dangerous. I warn you strongly not to abuse me. If you do I will prick you and pierce you like the prickly pear.
Lucas Mangope, speech to Braklaagte residents, 19 May 1989.

Introduction

During the 1970s, the National Party set in motion its plans for the creation and then consolidation of the Bophuthatswana homeland on the basis that its black inhabitants were culturally and politically homogeneous. All the land comprising the western Transvaal bushveld and the former Bechuanaland reserves was to be incorporated into the nascent 'state'. Apart from a few urban nodes, such as some townships around Pretoria and Brits, the homeland comprised predominantly rural people. This chapter examines how the occupants of Bophuthatswana were affected by and reacted to this development. From the outset, the period witnessed a good deal of civil conflict between the various African communities occupying the area and the Bophuthatswana state, as well as internal disputes within several *merafe*.

Like all the homelands, Bophuthatswana was created on the platform of what the government interpreted as 'traditional' leadership. Through elections, political parties and a presumed democratic process, the homelands were provided with the illusion of independence. As indicated in Chapter Four, like his father before him, Lucas Mangope was the government's preferred candidate to lead the Tswana people to independence and had been to some extent groomed to take over. This meant that he had to establish dominance over those rivals whom he saw as a threat to his authority. It was 'a particularly sore point with Mangope, as his own traditional authority over others was widely refuted, and he depended to a large extent on the support of the South African state'.[2] One of those who contested the elections to the Tswana Territorial Authority, the precursor to the Bophuthatswana Legislative Assembly, was the Kgatla *kgosi*, Tidimane Pilane. Pilane's conception of a Tswana homeland was that it should at a later stage amalgamate with other homelands. He claimed that his party, the National Seoposengwe Party, stood for African, as opposed to narrowly Tswana, unity, and espoused a form of 'federalism'. The National Seoposengwe Party wanted a full referendum to gauge the support for independence. In contrast, Mangope's Bophuthatswana National Party 'embraced a far narrower ethnic and chiefly constituency'.[3] The run-up to elections for full blown 'independence' was hampered not only by the National Seoposengwe Party's generally contrary view of the entire system but also by a constant power struggle within the Bophuthatswana National Party itself. In 1975, the Bophuthatswana National Party effectively split, leading Mangope to form a new party, the Bophuthatswana Democratic Party. His opponents threw their lot in with the National Seoposengwe Party. When fresh elections were held for the Legislative Assembly, Mangope won a clear majority and, with the appearance of legitimacy conferred through these elections, Bophuthatswana was led down the road to independence, an act which stripped over two million people of their South African citizenship. The move was not by any means fully endorsed by the residents of the future 'independent state'.

The National Seoposengwe Party, led by Pilane and, later, Victor Sefora, had strong support among the baKgatla and the baFokeng. Initially, this may have been based on a sense that Mangope lacked the status to be president of Bophuthatswana, but soon more material grounds based (as this account will show) on mining royalties formed the real factor in this contest. In the 1970s many opponents of the Bophuthatswana regime opted to work within the National Seoposengwe Party as a (legitimate) way of exposing the political fraudulence of the homelands. In truth, the National Seoposengwe Party never really found a coherent policy or set of underlying principles, probably because it was never its intention to seriously challenge for power in Bophuthatswana; moreover the evidence is incontrovertible that it had close contact and linked interests with the ANC in Botswana.[4] It was an amalgam of various people opposed to Mangope for different reasons, and lacked a sound electoral base. By the 1980s, it was a target of harassment by Mangope's security apparatus, and Sefora was arrested in 1979 and several times in the 1980s. There was therefore bad blood between Mangope's Bophuthatswana Democratic Party and the National Seoposengwe Party. This was to come to a head in the late 1980s.

FIGURE 16: Lucas Mangope on his farm near Motswedi, c.1975.

Source: National Archives of South Africa

Turmoil in the bushveld: The baKgatla ba Kgafela

In the 1980s, Mangope tried (and to an extent succeeded) to bring many powerful *merafe* in the bushveld and Bophuthatswana under his direct control. The first were the baKgatla ba Kgafela in the Pilanesberg/Rustenburg area. The Bophuthatswana government established the Sun City resort and the adjacent Pilanesberg game park in the early 1980s. Both generated considerable income and, in the case of the Pilanesberg, some worldwide acclaim. The stated ethos behind both projects, which were begun in the late 1970s, was that they should employ conservation to the benefit of the local population, using the jobs created and the income from hunting and entrance fees for social services. As we describe in more detail in the final chapter, both projects also generated some controversy, the Sun City resort being criticised as an island of pleasure in a sea of poverty. In addition, much of the land set aside for the Pilanesburg game reserve belonged to the baKgatla. The arrangement was that the baKgatla would be compensated with additional land for that incorporated into the park, an arrangement involving approximately 104 families, among them an *oorlam* (acculturated Dutch/Afrikaans speaking African) community who had lived on the farm Welgeval within the park's boundary since the 1860s.

Many Kgatla men, particularly those who stood to lose grazing land, were not satisfied with this arrangement, but when objections were raised and the matter stalled the Bophuthatswana government allegedly threatened the baKgatla that it would take away other land owned by the chiefdom. In a report tabled by an anthropologist from the University of the Witwatersrand, Jeremy Keenan, Chief Pilane attributed this threat to the fact that he was honorary president of the opposition National Seoposengwe Party. Kgatla spokespersons also claimed that the extra land set aside for compensation was taken over for the personal use, among others, of the Bophuthatswana minister of lands and rural development. The baKgatla subsequently threatened to sue Mangope for an amount of just over R13 million for inadequate compensation, following the loss of some 8 500 hectares of their land. Although the concept of the park was in itself probably of potential general benefit, the way in which Mangope steamrollered over the objections of the baKgatla illustrates once again his intolerance for the feelings of the citizens of his 'independent homeland'.[5]

Braklaagte and Leeuwfontein

One of the most sustained incidents of political resistance, matching the 1957 Hurutshe revolt, took place in the bushveld twin villages of Braklaagte and Leeuwfontein.[6] As we have noted above, the removal of these two 'black spots' in the 'Marico corridor' of white farms had coincided with the uprising of 1957. However, as the research of Kobus du Pisani and Arianna Lissoni has revealed, attempts had been made to deprive the villagers of a home long before this date.[7] The Braklaagte and Leeuwfontein troubles arose as a consequence of Bophuthatswana's determination to enforce the incorporation of the two villages into the homeland at all costs and without consultation with the communities. The announcement to reincorporate Braklaagte, a Hurutshe village some twenty-five kilometres north of Zeerust, was made in 1986.

The fear of being incorporated into the homeland arose from the villagers' anxiety that they would not be able to work in South Africa and that they would lose independent control over their land. Moreover, the incorporation issue was closely linked to a struggle over control of the two villages, with Mangope interfering most strongly to place his preferred candidate in control. There had been a long-standing dispute with the baHurutshe at Dinokana as to who was the rightful leader of the Braklaagte faction. The *kgosi* at Braklaagte, from approximately 1950, was John Lekoloane Sebogodi – however, he was considered to be a 'headman' who was substituting for Moitasilo Edwin Moiloa, the young son of the former deceased *kgosi*, George Moiloa. The enmity between Sebogodi and the Moiloa family at Dinokana went back to the first decade of the century, (see Chapter three) and worsened during the revolt of 1957-58 when the baHurutshe had been roughly divided into those who supported the state and those who opposed it. Sebogodi was clearly identified with the latter, while Godfrey Moiloa in Dinokana was identified with the body of traditional leaders who united under the authority of the Mangope family and who broadly opposed the resistance and, later, openly supported the bantustan policy. When Edwin found that the majority of the people at Braklaagte opposed his instalment as leader in the village he threw his lot in with Godfrey and supported the incorporation of the two villages into Bophuthatswana, thereby ensuring government backing for his appointment as *kgosi*.

The impetus for incorporating the villages into Bophuthatswana grew as consolidation plans for the homeland reached finality, and its mythical borders were delineated by apartheid planners in Pretoria. This happened to coincide with Sebogodi's impending retirement. Getting rid of Sebogodi and replacing him with a pro-Bophuthatswana *kgosi* was an important preliminary step in Braklaagte's incorporation – the problem, however, was that the majority of the

FIGURE 17: Father and son, Lekoloane and Pupsey Sebogodi, ruling family. Braklaagte c. 1989.

Source: Gille de Vlieg

community totally rejected Edwin's claim to chieftainship and gave their support to Pupsey Sebogodi, the son of Lekoloane.

The South African government was nevertheless determined to incorporate the area, and set a date for the end of 1988. The residents were to fall under the authority of Godfrey Moiloa and their land would be held in trust for them by Mangope as president of Bophuthatswana. The community objected and mounted a campaign of petitions and letters to the press, and eventually Pupsey Sebogodi went with a delegation to see the minister of constitutional affairs, Gerrit Viljoen, believing he might be sympathetic to their cause. They presented a memorandum setting out their historical separation from the house of Moiloa at Dinokana, re-affirming that they were a longstanding independent community, and stating that they wanted to remain South African citizens with the resulting economic privileges. Two weeks later, however, Braklaagte was incorporated, along with the Leeuwfontein, which had at no time prior to this been mentioned in the proposed incorporation. The Braklaagte community immediately took the matter to court, giving reasons for their objections to incorporation, and requesting the South African government to provide reasons why incorporation was valid. The Bophuthatswana judge, however, ruled in March 1989, that the proclamation was valid. The matter was taken on appeal.

After this, between April and July, the two villages were subject to severe harassment by the Bophuthatswana security forces. In March a large contingent of Bophuthatswana police and army set up camp in Braklaagte. Scholars mounted a series of school boycotts for which they were arrested and assaulted; Sebogodi was detained under the Bophuthatswana Internal Security Act; and the establishment of a police camp in Leeuwfontein sparked off repeated confrontations between the police and the residents. Sebogodi and another sixty people were charged with arson and public violence. Mangope addressed residents on 19 June, threatening them with the citation at the front of this chapter: 'Beware that Bophuthatswana is like a prickly pear.'

On 1 July (subsequently referred to as 'black Saturday'), residents planning a meeting were ordered to disperse by the Bophuthatswana police who had received information that 'agitators' had arrived from the Witwatersrand. Initially, a police vehicle moved into the crowd, spraying teargas and rubber bullets. In the ensuing confusion, nine policemen and two residents were killed – four of the policemen had been trapped in a burning vehicle. The event thrust the two communities onto the national stage, leading to widespread comment that was generally critical of the state's determination to enforce incorporation against the wishes of the residents that had been the basic cause of the tragic loss of lives. State retribution was swift and brutal. Claiming that the 'cold blooded murders' were being planned in Johannesburg, the police sealed off the villages and the perpetrators of the killings were hunted down, on occasion with the aid of helicopters. One hundred and thirty six people were detained and held; thousands of residents fled or took refuge in the nearby hills or on the farm of a neighbouring white farmer, Paul van der Merwe. The lawyers that the community had engaged (Bell, Dewar and Hall from Johannesburg) were repeatedly refused access to their clients. Even when they applied successfully to the Supreme Court in Mmabatho for such access, their lawyer James Sutherland (who also acted for the baFokeng) was refused a visa to enter Bophuthatswana.

The press was banned from the two trouble spots, and the Black Sash and the Transvaal Rural Action Committee, which had been active as a support group in the area, were banned in Bophuthatswana. The policemen who lost their lives were buried with full military honours. Ironically, with feelings still running high, Mangope went to London to host a tea party for Conservative MPs, to convince them to recognise the homeland. A pervasive climate of oppression hung over the district until the end of the year.[8] During this time, the South African government refused to intervene, the ambassador to Bophuthatswana, Dr L Kotze, claiming that the communities were 'children of Bophuthatswana'.[9] A prolonged trial took place at Rooigrond, near Mafikeng, and over thirty-one people were ultimately charged with murder. A few men were found guilty of public violence and two of murder, for which they were given life terms. In June 1990, charges against most of those accused were dropped.

Braklaagte residents joined the wider movement for the prevention of homeland incorporation into South Africa, and Pupsey Sebogodi was elected chairperson of the Anti-Bop Coordinating Committee which campaigned for the immediate disbandment of the homelands.[10] At an international level, the city of Atlanta in the US undertook a campaign of support for Braklaagte, even entering into a twinning agreement with the remote village in South Africa's bushveld. However, despite such developments, the measure of trepidation and mistrust was so intense that only in 1991 did the majority of villagers dare to return to their homes.

Ultimately, and similarly to the conclusion of the Bophuthatswana-baFokeng dispute described below, it was unfolding events in South Africa that decided the fate of the people of Braklaagte and Leeuwfontein. In 1990, FW de Klerk, the state president, announced the unbanning of the ANC and other liberation groups, and the release of Nelson Mandela. The writing was on the wall for the homelands, although Mangope resisted re-incorporation into South Africa until the very last moment, when he was overthrown in the coup of 1994. With the inevitability of the collapse of the homeland system there was little further protest from the two villages, and the matter subsided into a kind of limbo. Interestingly, a significant number of ANC branches sprang up in Braklaagte although it remained *de facto* banned in Bophuthatswana itself, as South African political organisations were in law not allowed to operate there. The deaths and suffering of the residents of the villages, and the loss of life of the Bophuthatswana policemen, so soon before the restoration of democracy in the country, was tragic and unnecessary but the resistance mounted by the villagers of Braklaagte and Leeuwfontein nevertheless ranks high in the annals of rural resistance to the apartheid system. As Du Pisani's study concludes, the two communities 'took an active role in their own history and did not let themselves be pushed around by state officials [using] every opportunity afforded to them by the weakness of the apartheid system to resist their relocation ... an achievement they kept up for eighty years.'[11]

Mathopestad and Mogopa

Much the same could be said for the baKubung of Mathopestad and the baKwena of Mogopa. The former, it will be recalled, had left the main section of the baKubung ba Ratheo near Ventersdorp in the late nineteenth century. They had been earmarked for removal into the nascent Bophuthatswana since 1968, but resisted by every means possible, mainly because they had had the opportunity to assess the dire consequences of the removal of the baKubung ba Monnakgotla from Molotestad to Ledig. The proposed resettlement of the Mathopestad people was to be at a place called Ondersterpoort, near Ledig. To highlight their situation, the Mathopestad residents developed a media strategy which led to a highly publicised visit by Senator Edward Kennedy to Mathopestad in January 1985. Kennedy called the forced removals policy 'inhumane and indecent', and met with South Africa's minister of foreign affairs, Pik Botha, to tell him so. Botha, in his trademark whimsical style, informed Kennedy that the villagers were being removed for 'medical and hygienic reasons'.[12]

The state's tactics, as they had been in the case of the baKubung at Molotestad, were to intervene in the leadership structures of the community in an attempt to divide them and lure a chief into accepting the supposed benefits of moving. Cyprian Mathope, the long-term *kgosi* (1936-1977), refused to cooperate with the authorities, but his son Arthur showed signs of collaboration. In March 1982 the *khuduthamaga* censured him following information that he had been secretly meeting the native commissioner in Ventersdorp, and he died under mysterious circumstances a few months later.[13] The native commissioner managed to negotiate with the baKubung to secure the appointment of Arthur's uncle, Solomon, but when he also proved obdurate to the government's plans, the native commissioner intervened again, suggesting that Arthur's son should become the *kgosi*. The community stood behind Solomon. Possibly because of Kennedy's intervention, the

FIGURE 17: Mathopestad women bringing in beer following the announcement that the removal would be halted.

Source: Gille de Vlieg

'black spot' of Mathopestad was never removed, one of the very few areas not forcefully relocated under the apartheid system.

The baKwena ba Mogopa owned two farms near Ventersdorp which they had occupied from the time of the 1913 Natives Land Act. On one of them was their main village, Mogopa. Their welfare and general security was threatened with the appointment of Jacob More as chief in 1978. A former policeman, More was accused by many community members of corruption and an authoritarian style of leadership, and in 1981 the community overwhelmingly voted to remove him and his councillors from office. However, the native affairs commissioner, a Mr de Villiers, stepped in and insisted that More was the legitimate chief.

There was a reason for this that the villagers did not know at the time. Mogopa was destined for removal, and the government wanted a compliant *kgosi* in power who would do its bidding. In February 1982, it was announced to an astonished gathering of the baKwena that they would be removed to a place called Pachsdraai in Bophuthatswana – Jacob More had colluded in this plan to remove the residents. In June, about a hundred families moved in dribs and drabs to Pachsdraai (More himself had been given a former white-owned farm to reside on in return for having backed the government's plans).

The remainder of the community refused to budge, for which they suffered greatly. There were no schools or water provision, and buses to Ventersdorp stopped running. People were in daily fear that their houses would be demolished. With the assistance of the Black Sash the Mogopa resisters hired a lawyer who obtained a guarantee from the minister of cooperation and development that no houses would be demolished – despite the fact that a bulldozer was parked in the village next to the demolished school. It was a short-lived triumph, for a few weeks later the state president, using an order in terms of the Black Administration Act of 1927 ordered the removal of the remaining residents by November 1983 and it was arranged that government trucks would come to move the possessions of the remaining villagers.

FIGURE 18: Walking past ruins – Mogopa, December 1983.

Source: Paul Weinberg

The community resolved to refuse to move and further (largely unsuccessful) negotiations were entered into with the minister. A stay of execution of the eviction order also failed. The Black Sash gave considerable publicity to the Mogopa case, ably assisted by Helen Suzman, the sole Progressive Party parliamentary representative, who was in New York at the time and personally approached Dr Chester Crocker, the assistant secretary of state for African affairs. Pik Botha who was on a diplomatic charm offensive tour of Europe at the time was also clearly embarrassed by the incident as Mogopa now hit the international headlines and when the deadline for removal came, the (then) Bishop Tutu conducted an all-night vigil with a host of international media in attention. The deadline came and went. No trucks arrived. The state backed down and the Mogopa people remained in their village but, with no schools or services, a number of families slowly did go to Pachsdraai.[14]

Other bushveld communities

Mangope's ongoing determination to crush, and/or replace rurally-based opposition, particularly among non-Tswana chiefdoms, continued into the 1980s. First to feel the brunt was *Kgosi* Bosman Noah Ramokoka of the baPhalane who lived at Ramamkokstad in the eastern part of the Pilanesburg, about fifty kilometres north of Rustenburg. He quarrelled with Mangope and was deposed, despite a commission of inquiry's recommendation that he retain his position. True to form, Mangope appointed an acting chief and the community was reportedly in chaos in 1984.[15]

The baTlhaping

Mangope had alienated the baTlhaping ba ga Phuduhucwana in Taung from at least 1983, when he ordered the Bophuthatswana Defence Force to shoot

FIGURE 19: Mogopa resident moves his personal belongings by wagon.

Source: Paul Weinberg

'surplus' donkeys in all districts of Bophuthatswana. This was a year of crippling drought, and there was insufficient grazing for cattle and other domestic live-stock, let alone donkeys, but it was a severe miscalculation on his part, for don-keys, as Nancy Jacobs has indicated, had become almost indispensable to rural dwellers in Kuruman and indeed throughout the North West Province from the beginning of the twentieth century. They were used for transportation and were cheap to maintain as they could survive on a poor quality diet.[16] This has remained so until the present, as a drive around the rural areas of the province will testify. According to Jacobs, the massacre did not match the potential harm the donkeys presented, and may have been 'politically motivated, intended to terrorise people and pre-empt opposition',[17] but it could also have been a case of sheer panic and political misdirection. Nevertheless, it took on a bloodthirstiness that shocked the local inhabitants and was considered by most rural Africans to be a 'serious moral transgression'. [18]

Ill-feeling was intensified in 1987 when the ruling *kgosi*, Scotch Mankurwane, died, and a regent, Samuel Mankurwane was appointed by the chiefdom to act for his minor son. Samuel was a graduate of the University of Natal and a supporter (if not an affiliated member) of the United Democratic Front (UDF) which regarded the bantustan system as a fraud. He was patently not to Mangope's lik-ing, and in true Mangope fashion was deposed as chief. A new regent, Stephen Molale was appointed in his place. Samuel's supporters took the matter to court and briefly managed to have him restored, but Mangope amended the Traditional Authorities Act to give him the legislative powers necessary for Samuel's deposi-tion. (Later, he was detained and, according to his own testimony, tortured, ridi-culed and harangued by the the notoriously intolerant figure of Major General Seleke, then acting minister of justice in the homeland.)[19] In the meantime Molale took charge of Tlhaping affairs against the desires of a majority of the *morafe*. In 1990 Samuel Mankurwane took the well-trodden route to exile in Botswana. His supporters kept the issue of the chief's plight in the public eye by holding protests at the Bophuthatswana Consulate in Johannesburg.[20] However, like many of the disputes that involved the Bophuthatswana government and its citizens, the mat-ter was only really resolved after 1994 when the bantustan disappeared.

The baFokeng

Finally, we give consideration to the baFokeng.[21] They were the most crucial, because their dispute with Mangope was linked to a battle for the ownership of mining rights and control of assets from mining on centre stage. It was also the area where probably the most traumatic, and certainly the most prolonged, epi-sode of resistance to Mangope's rule took place. This occurred essentially in two phases: the 1980s (when the conflict was mainly on a political basis) and the 1990s (when it was linked to a legal contest over the rights to royalties on platinum). As will be seen, these were not entirely discrete phases, as from the start mineral profits were by no means peripheral to the contest.

The fallout between the baFokeng and Mangope, and the concomitant struggle over royalties, is a saga that has been recounted elsewhere.[22] Essentially, the Fokeng *kgosi*, Edward Lebone Molotlegi, threatened to secede from Bophuthatswana in 1983, although he had flirted with homeland politics himself in earlier years. Mangope promptly declared a state of emergency in the capital, Phokeng, and went on to establish a commission of inquiry into Lebone's alleged malpractice.[22] At the core of the quarrel was probably Mangope's desire to divert royalties from the baFokeng to the coffers of Bophuthatswana, and Lebone's attitude provided him with an opportunity to take charge of their affairs.

An even better opening for Bophuthatswana's intervention was created by the attempted coup of 1988 (the coup leader, Rocky Malebane-Metsing, hailed from Phokeng, which was an opposition stronghold). Lebone was detained and accused of supporting the coup. Upon his release he decided to go into 'exile' in South Africa and, later, Botswana. With the chiefship vacant, Mangope as president had the right to appoint an acting *kgosi*, although he was obliged to consult with the baFokeng, and he appointed Lebone's brother, Mokgworo George Molotlegi, against the wishes of Lebone and a majority of the Fokeng *lekgotla*, who favoured his uncle, Cecil Tumagole. What prompted 'acting chief George', as some disparagingly termed him, to take up Mangope's offer is a matter for speculation, but apparently there was bad blood between the brothers. Whatever the case, George essentially became Mangope's pawn, which enabled him to interfere increasingly in Fokeng affairs.

In March 1989, the Fokeng supporters of Lebone, not for the first or last time, took the long litigious route to resolve the impasse with Mangope. The legal firm of Bell, Dewar and Hall bought an urgent application to have George's appointment set aside and to have Cecil Tumagole installed as Lebone's deputy. It was contended that Mangope had transgressed Fokeng law and custom in making the appointment. However, in a case heard before Bophuthatswana judges Smith and Friedman, the court found that Mangope had acted legally in accordance with the Traditional Authorities Act of the homeland. The Fokeng lawyers disagreed and took the matter up on appeal.

While this lengthy case was being played out, Mangope began to hound Lebone's wife, Semane. With her husband in exile, she was now the visible object of his thwarted ambitions to take full control of the baFokeng, and although she had significant support from the chiefdom and from within South Africa she was detained in prison along with members of an innocuous self-help organisation called the Bafokeng Women's Club. Her house was searched several times, and then she was placed under house arrest. The Fokeng lawyers then successfully interdicted Mangope, who was both president and minister of law and order, from harassing Semane and preventing the baFokeng from meeting to discuss their problems. However, on appeal the Bophuthatswana High Court overturned the interdict. In 1989, Semane Molotlegi's visa to live in Bophuthatswana was not renewed and she had little option but to go into 'exile' herself, ending up in Botswana with her husband.

This political battle was part and parcel of Mangope's concern to benefit from the burgeoning mining industry in Bophuthatswana. In 1921, an ore-bearing reef containing metals of the platinum group, the Merensky reef, was discovered in the bushveld complex close to Rustenburg. A large portion of the reef was situated on Fokeng land which had been acquired mostly in the nineteenth century, largely due to the efforts of Mokgatle Thethe. By the turn of the century, twenty-two properties were registered in the name of the baFokeng, and held in trust for them, first by the HMS missionaries and after 1881 by a state functionary.[24] In the next century another eleven farms were acquired. The baFokeng were aware of the potential of mineral deposits, but the mining of platinum is a costly and complex affair, and is driven by external factors such as the demand for the metal. Up to this time the baFokeng remained relatively poor, and many men and women were obliged to seek work in nearby Rustenburg, or on the Reef, as the work of Belinda Bozzoli and others has shown.[25]

By 1968, however, the combined assets of Impala Platinum Holdings (Implats) and Anglo-American Platinum (Amplats) were such that mining could begin in earnest. In those times, the owner of the mineral rights was severely disadvantaged, as the mining industry possessed most of the technical and geological information on mining and could therefore easily dupe the owners when agreements were being negotiated. The odds were stacked against the owners of the mineral rights in other ways as well. For example, the granting of prospecting rights did not set out in law to what the baFokeng were entitled – it was simply a matter of negotiation, and when applying for a mining lease the mining companies did not need to obtain the permission of the baFokeng. In addition, there was no basis on which royalties were payable – often they were just an improvement on earlier offers. The baFokeng did not get their first royalty until 1978. 'It seems breathtaking,' they later observed, 'that … the baFokeng community, the owner of the ore reserves, is referred to for the first time in Impala's Annual Report published in 1988, twenty years after Impala commenced mining operations.'[26]

From 1985 the baFokeng began to question the nature of this relationship with Impala. Their first tactic was to try to commence mining by themselves, so they approached Impala for more information on prospecting and mineral deposits. This alarmed Implats and it approached the Bophuthatswana government to see if there was any means of forestalling the Fokeng request; Implats was also trying to get a mining lease over a potentially rich vein of ore known as the 'deeps' and wanted to know whether the homeland could assist. The request for the intervention of the Bophuthatswana regime was fortuitous for Mangope, as it drew Bophuthatswana into an alliance with Impala and introduced the question of trusteeship over the mineral-rich land that comprised a significant portion of the famed Merensky reef. Edward Molotlegi's deposition as *kgosi*, and Mangope's determination to appoint in his place someone amenable to the government, should be seen in the context of this development. The question of trusteeship over Fokeng lands now became a crucial issue, Mangope claiming that as president he was the rightful trustee.

There followed nearly ten years of vexatious legal proceedings that hinged around the 'improper' relationship between Impala and Bophuthatswana, and Mangope's right to act as 'trustee' of Fokeng affairs in the course of which the independence of the Bophuthatswana judiciary was in serious doubt. In fact, several 'pertinent examples of the lack of independence in the Bophuthatswana judiciary' specifically Mangope's powers to appoint and fire judges, were raised by the Fokeng lawyers to show that there had been an 'improper' relationship between Impala and the then president.[27] It was only in 1994 after the unbanning of the ANC in South Africa, and the ousting of Mangope in a popular coup in March of that year, that the scene was set for the baFokeng to resume in earnest their fight for royalty rights. Lebone Molotlegi returned from exile, the Supreme Court of Bophuthatswana overturned George's appointment, and the Fokeng lawyers entered into fresh negotiations with Impala. Inexplicably, given the markedly transformed political landscape, Impala continued to challenge the status, or *locus standi*, of the baFokeng as legal owners of the land. Negotiations again broke down in April 1997, despite the attempts of both legal teams to find a compromise; this meant that the Fokeng legal team had to go to court again to re-embark on steps to nullify Impala's right to mine on their land. When faced with the weight of evidence against his earlier decision, Judge Friedman had little option but to find in favour of the baFokeng, in the process repudiating his earlier opinions. He accepted the argument advanced by the baFokeng lawyers that it would have been a miscarriage of justice if the baFokeng had not been granted *locus standi*. This paved the way for a settlement that was in fact beneficial to both sides – with Impala's right to mine on Fokeng land now ensured, it cleared the way for the company to enter into a merger with Lonrho that it had been negotiating for several years, and led to huge increases in its share prices. From the baFokeng position, the royalty payable would be increased to 22 per cent of the taxable income in respect of the key mining areas, up from approximately 15 per cent in the past. The baFokeng also received one million shares in Implats and had the right to nominate one person to the board of Impala Platinum. In addition, various arrangements were put in place to reduce the potential for further conflict. In cash terms, the increase in royalties amounted to R100 million annually, although this was obviously likely to fluctuate from year to year. The press hailed the generous terms of the settlement, and the baFokeng were fêted as 'the richest tribe in Africa'.[28]

Thus ten years of costly and acrimonious litigation was brought to an end. In retrospect, one is forced to wonder why this should have been the case, given that the settlement was mutually beneficial. Possibly it became a matter of pride that Impala should win the war of legal attrition. Despite the enormous wastage of resources in the battle between Impala and the baFokeng, one positive point emerged: the litigation 'changed the rules' (in the words of the baFokeng attorney, James Sutherland) between the owner of the mineral rights on the one hand and mining institutions on the other, a relationship that had been vastly skewed in favour of the mining companies.[29] No longer could mining companies exploit traditional communities such as the baFokeng in ways that they had done in the past. The effect of this unexpected and fabulous windfall on the baFokeng is the subject of the next section of this book. As the account above clearly indicates, the contest for economic independence by the baFokeng was linked to a fight for political independence from the Bophuthatswana state under Lucas Mangope.

A somewhat different slant on the shifting three-way contest between Mangope, Impala and the Fokeng chieftaincy is provided by Gavin Capps. He introduces what is really a fourth 'player' in the equation, organised labour, in the form of the National Union of Mine Workers (NUM). The labour union began to gain a foothold among the black mining force in the Rustenburg area from about mid-1990 despite the fact that Bophuthatswana's restrictive labour legislation forbade any 'foreign' trade unions from operating in the homeland. In 1991, workers at Impala's mine shafts went on strike, catching both the NUM and Impala management off-guard. The NUM quickly picked up on the strike's momentum, and sent in organisers to assist the workers and recruit among them. Impala was forced to recognise NUM's presence and concede operating rights to it in the homeland, which antagonised the Mangope regime. Impala was thus placed in an invidious position – it needed to end the strike, stabilise labour relations and resume production, yet was anxious not to offend Mangope.

The strike thus fractured the alliance between the two, obliging Impala to seek reconciliation with the baFokeng. In this analysis, emphasis is placed on the economic struggles of the organised working class in Bophuthatswana in breaking the deadlock, rather than on legal or political contests. However, in Capp's analysis the deal with Impala strengthened the position of the Fokeng chieftaincy by uncomplicatedly transferring trusteeship of land within the baFokeng area into its hands. As all of this occurred during the delicate period of the country's transition to democracy, Capps asserts that by default the state had abandoned its fiduciary and oversight role to African communities in the baFokeng area of jurisdiction.[30] His interpretation is consistent with his view of the Fokeng ruling elite as land-grabbers who exploited anomalies in South Africa's discriminatory land ownership laws to effectively 'privatise' farms in the baFokeng area at the expense of smaller but independent landowners.

A similar picture is painted by the labour historian R.M. Reddy who, in an examination of the labour unrest of that time, argues that it 'evolved from an industrial dispute into a highly politicised engagement that posed a serious direct challenge to the Mangope regime', and which played an 'important, albeit over-looked, role in bringing about the downfall of the homeland'.[31]

ENDNOTES

1   See *Tsala ea Batho*, 31 May, 1913. The report makes reference to Mangope being a leading chief in the SANNC in the Zeerust region.

2   B. Mbenga and A. Manson, 'Resistance and Repression in the Bantustans; Bophuthatswana', p. 784, in South African Democracy Education Trust, (SADET), *The Road to Democracy in South Africa*, vol. 2, 1970-1980. (Pretoria, 2006). Other sections of this chapter are drawn from this same source. The argument is also presented in M. Lawrence and A. Manson, '"The Dog of the Boers": The Rise and Fall of Mangope in Bophuthatswana', *Journal of Southern African Studies*, vol. 20, no. 3 (1994).

3   Mbenga and Manson, 'Bophuthatswana', p. 787.

4   We draw this conclusion from an examination of the Oliver Tambo Collected Papers, microfilm A2651, Historical and Literary Papers Collection, Papers, William Cullen Library, University of the Witwatersrand, which provide evidence of direct communication over strategy between the two.

5   Bell, Dewar and Hall, See *Weekly Mail*, 12 July 1985.'Animal Paradise-No Cows Please', *Sunday Tribune*, 14 July 1985.

6   A recent and full account of the Braklaagte and Leeuwfontein 'troubles' has depicted them as a frontier war, and a struggle for land. See K. du Pisani, *The Last Frontier War: Braklaagte and the Struggle for Land*, (Rosenberg Publishers and UNISA Press, Pretoria, 2009). Our view is that this was a prolonged act of resistance to maintain political autonomy, though access to land formed part of this objective.

7   Lissoni, 'The Bahurutshe Chieftaincy', p. 15.

8   This summary of events is taken from the Black Sash, '"Grasping the Prickly Pear": The Bophuthatswana Story, (Black Sash, Johannesburg 1990),and the Black Sash, 'Fighting the Drawings of a Pen', (Johannesburg 1991). See also

Business Day, 2/8/1989; *Sunday Time*s, 9/7/1989; *Star*, 12/7/1989; *Weekly Mail*, 14 to 20 July 1989. A more detailed version is contained in Du Pisani, *The Last Frontier War*.

9   *New Nation*, 19 May 1989.

10   See *New Nation*, 20-26 October 1989.

11   Du Pisani, *The Last Fronteir War*, p. 237.

12   *Los Angeles Times*, 7 January 1985.

13   Dugmore, 'The Bakubung of the Western Transvaal', pp. 104-105. Some sources report that he died after being administered 'medicine' by members of the *morafe*.

14   This account is taken from "Mogopa Rebuilds", A Story of Resistance', *The Blach Sash* February 1984 2-6. The incident received quite wide coverage in the local press.

15   *Sunday Times*, 12 January 1985.

16   Jacobs, *Environment, Power and Injustice*, pp. 123-125.

17   Op. cit., p. 201.

18   N. J. Jacobs, 'The Great Bophuthtswana Donkey Massacre: Discourse on the Ass and the Politics of Class and Grass', *American Historical Review*, 108 (1997), p. 18.

19   See SADET Oral History Archive, Interview with S. M. Mankurwane, conducted by B. Mbenga, Taung, 15 April 2004.

20   See also A. Manson and B. Mbenga, 'Bophuthatswana in the 1980s and the UDF in the Western Transvaal', SADET, *The Road to Democracy*, vol. 4, 1980-1990 part 1, pp. 674-676.

21   A lot of this discussion is taken from Mbenga and Manson, *People of the Dew*, though some new material has been added.

22   A. Manson and B. Mbenga, '"The Richest Tribe in Africa": Platinum Mining and the BaFokeng in South Africa's North-West Province, 1965-1999,' *Journal of Southern African Studies* vol. 29 (2003), pp. 25-47. Much of the information was obtained from James Sutherland, the lawyer acting for the BaFokeng, in a number of interviews held in August 1999. See also Mbenga and Manson, *People of the Dew*, Chapter 7.

22   *Mafikeng Mail*, 20 May 1983.

23   For more on this see J. Bergh, '"We must never forget where we come from": The BaFokeng and their land in the nineteenth-century Transvaal', *History in Africa*, 32 (2005).

24   B. Bozzoli with M. Nkotsoe, *Women of Phokeng: Consciousness, Life Strategy and Migrancy in South Africa, 1900–1983* (Johannesburg: Ravan, 1991).

25   Cited in Manson and Mbenga, 'The Richest Tribe in Africa' p. 29 from *Sunday Times*, 3 August, 1999.

26   Mbenga and Manson, *People of the Dew*, p. 150.

27   *Business Day*, 9 February 1999.

28   *See Manson and Mbenga*, in 'Richest Tribe in Africa', p. 46, from interview with James Sutherland, 4 August 1999.

29   This is expounded in Capps, 'Tribal-Landed Property', Chapter 7.

30   R.M. Reddy, '"Darlings No More": Tswana Mineworkers and Labour Unrest on the Platinum Mines of Bophuthatswana', B.A. (Hons) essay, University of the Witwatersrand, 2011.

Modernity in the bushveld:
Mining, national parks and casinos

The 'immeasurable wooded plain' (Delegorgue's 1843 depiction of the bush-veld) which we mentioned in the Introduction was radically transformed over the course of the ensuing 150 years. Commercial agriculture had long been practised along the Marico valley, the undulating lands between the towns of Zeerust and Swartruggens, and the fertile plains surrounding Rustenburg. The bushveld had been tamed, although in pockets it still retains its rustic charm as captured in the quixotic works of the novelist Herman Charles Bosman. However, a number of new and salient features become visible. The first are the many mines that scar the countryside around Rustenburg in the controversial 'platinum belt' of South Africa's mining industry, the site of long-running labour and social unrest and violence that ultimately led to the Marikana shootings. The second are the numerous game parks and reserves that form the basis for the North West Province's tourist industry. The two major game reserves are the Madikwe Park to the west, adjacent to the border with Botswana, and in the east the Pilanesberg (and the adjoining Sun City entertainment centre), which occupies all of the fertile volcanic terrain of the Pilanesberg crater. There has even been talk of joining up the two to form an extensive conservation reserve.

Whereas many black people still live under the nominal control of tribal authorities (see Map 4), increasingly, especially in the mining districts around Rustenburg, large numbers of workers from other parts of the country have set up homes. As the following account makes clear, economic success and advancement has had the effect of heightening ethnic consciousness, leading in turn to even more intensified power struggles between factions of royal families, between sub-ethnic factions, and between commoners and the traditional holders of power. These ongoing transformational developments, which now define the political, economic and environmental face of much of the wider Rustenburg region, are discussed below.

Mining and 'traditional communities': The BaFokeng Inc, 'the richest tribe in Africa'

How the baFokeng successfully obtained an equitable share of the royalties from Impala Platinum has been traced in the preceding chapter. However, in the new millennium, and with the worldwide platinum boom of 1996-2008, the baFokeng were able to emerge as major corporate players in the mining industry.[1] Their affairs were and still are largely conducted under the auspices of a Royal Bafokeng Administration, headed by the current *kgosi* (now termed king), Leruo Molotlegi of what is now called the Royal Bafokeng Nation. The Royal Bafokeng Administration has subsequently terminated the royalty agreements previously negotiated with the mining houses and embarked on a policy of direct investment in them. Thus, in 2007 royalties with Impala Platinum were exchanged 'for an issue of 9.4 million shares (worth R10.6 billion at the time) which made the Royal Bafokeng Nation the largest single shareholder in the world's second largest platinum producing company [Impala Platinum Holdings Ltd]'.[2] The Royal Bafokeng Nation's holding company held investments worth R33.5 billion in 2007, most of it derived from mineral and mining rights. In 2008 the Royal Bafokeng Administration announced further mining developments involving Impala's planned extension of its lease areas and the sinking of at least three more shafts.

The baFokeng have benefited from South Africa's black economic empowerment (BEE) policy, becoming, in the words of one academic, 'South Africa's empowerment partner of choice', clinching deals with the mobile phone giant Vodacom and the Thebe Investment Corporation.[3] Many more of their ventures could be added to this list: suffice it to say that this 'hugely complex array of investments' and the sophistication of their financial and investment arms have combined to create what the Comaroffs have identified as a prime example of 'Ethnicity Inc', the 'melding of the corporate and the cultural ... under the sign of the commodity'.[4] The baFokeng have also adopted the legal status of a *universitas persona*, a nonprofit organisation, operating in the interests of a community, and thus are exempt from tax earned on cash in the bank, totalling R5 billion in 2007.[5] (The South African Revenue Service has so far been unable to redefine the tax status of the baFokeng, despite its best efforts.)

The baFokeng have emerged as a distinct corporate entity whilst remaining a 'traditional' community, thus 'melding corporatisation with traditional governance'.[6] Furthermore, the modern Fokeng community combines 'nation building with business building' in a seamless interwoven pattern.[7] This ability to combine traditionalism with modernity has intrigued those studying the baFokeng.

As a result of this massive investment and economic expansion the Royal Bafokeng Nation has been able to establish a visible and positive social profile. The Royal Bafokeng Sports Palace, part of the Royal Bafokeng Holdings, was an official venue for the 2010 FIFA World Cup. Royal Bafokeng Holdings also has a stake in the Premier Soccer League club Platinum Stars and the Platinum Leopards rugby team, the first black-owned rugby franchise in the country. It also had, until 2009, a stake in the Black Tie Ensemble, a well-known operatic group. This accumulation of capital has led to generally (though some would dispute it) improved service delivery in the Fokeng region and massive infrastructural development. In 2007, a Royal Bafokeng Institute was established to improve the quality of education for the baFokeng, and the Lebone II College of the Royal Bafokeng was completed in 2010. The baFokeng have been remarkably successful in merging their operations with broader national interests, for example, in the construction of highways, the Bafokeng Sports Palace and investment in other ventures that benefit a wider community.

The Royal Bafokeng Administration, however, has had to deal with an increasing number of complex issues that arise from several factors. The first was hardships caused by a massive in-migration of labour from all over southern Africa into Fokeng areas, coupled with the appearance of large numbers of informal workers subsisting off the earnings of the miners. These immigrants are close to matching the baFokeng numerically – recent surveys place the population of the Fokeng areas at 142 000 of whom just over two-thirds are classed as Mofokeng.[8] This raises the question of how and with whom Fokeng wealth will be shared, and which localities will be targeted for development. As in most modern South African traditional polities, it is almost impossible to speak of a pure baFokeng ethnicity. In short, is Fokeng wealth being shared with those who do not belong to the baFokeng? A discernible sense of entitlement on the part of many Mofokeng presents a worrying trend even to those who represent them.[9]

Other critics believe the Fokeng wealth has not 'trickled down' to the wider community, and believe that most Mofokeng do not understand the complex financial and business deals the RBA engages in, thus creating an 'information gap' between it and the local people. The political economist Somwabile Mnwaba, who has examined community relations with the traditional leadership of the Fokeng (and baKgatla ba Kgafela) concluded that 'the interface between community participation is fraught with tokenistic participation, elite-targeted grassroots anger and local tensions'.[10]

In a previous chapter we outlined how individuals or minority groups have attempted to mount legal challenges to the exclusive ownership (or privatisation) of land by the baTswana ruling elites such as the baFokeng. In 2010, the Royal Bafokeng Administration went to court in the name of the current *kgosi* to terminate state trusteeship of over fifty-one farms held in trust for them. A number of smaller communities and private families, however, challenged the application on the grounds that the baFokeng were the not the only, or original, owners, and claiming that the properties had been acquired by their forefathers, either entirely or through having made significant contributions towards the purchase price.[11] This is a vexed issue, for there is evidence to support an alternative argument: namely that these private families have a history of interaction with the Royal Bafokeng Administration and have over decades become enmeshed with Fokeng society.

Another problem facing the Royal Bafokeng Nation relates to the kind of relationship it has with the Rustenburg municipality. Will it offer services independently from the local service provider or will it effectively become a state within a state? How will problems of overlapping authority, responsibility and jurisdiction be resolved? Former president Nelson Mandela remarked on this in 1995 on the occasion of the funeral of the late *Kgosi* Edward Lebone Molotlegi, when he warned that the establishment of 'democratic local authorities in rural areas needs to be handled with great sensitivity'.[12] Yet another problem derives from the environmental impact of mining, such as blast damage and pollution in and around Phokeng, a dangerous situation for all residents in the mining belt. Some people have had to be moved from their villages to make way for mining operations. Many have complained about receiving inadequate compensation, about the loss of grazing and ploughing lands, and about the provision of inadequate or inferior housing at their new sites.[13]

Those who endorse the legitimacy of the Royal Bafokeng Nation point out that their critics have not acknowledged either the extent of support the traditional ruling family commands or recent attempts to distribute resources in a meaningful manner. Nor have they acknowledged the measures taken to manage the effects of mining. Too often, the administrative efficiency of the Royal Bafokeng Nation and its financial capability have tended to be overshadowed by negative perceptions regarding elitism and the failure of levels of household income to rise. Supporters of the baFokeng chieftaincy point out that the benefits of investments from platinum revenues are not targeted at 'the individual or household level',[14] but are ploughed into communal programmes (for example health, education, environmental management) that will bear long-term benefit.

The Royal Bafokeng Nation argues that its critics have tended to hone in on the known 'hot-spots' where most of the mining has taken place. The particular dynamics at work in an area of rapid transformation: the in-migration of thousands of migrant workers; large-scale and rapid mining developments; relocations of communities, and so on, create conditions for dissatisfaction.

The baPo ba Mogale and Lonrho

The baPo, claiming about 75 000 followers, live in a cluster of seven villages in the Marikana district, halfway between Rustenburg and the mining centre of Brits, a distance of about 120 kilometres. One of these villages is Marikana, where the shooting of the miners (mentioned in the Introduction) occurred (see map 5). The baPo authorities entered into a notarial mineral lease agreement with JD Transvaal in 1969 which was later converted into a thirty year lease. At that time they were paid R2 000 per annum for royalties.

BaPo difficulties regarding the payment of royalties date back to the Mangope period. As he did with the baFokeng, Mangope established himself as trustee of the chiefdom and placed all their assets in what became known as the 'D' (development) account.[15] This slight was compounded by the fact that Mangope interfered in the internal affairs of the baPo: when the regent, Fred Mogale, died in 1982, the royal family should have appointed a successor. The oldest child in the first hut was a woman; consideration should have been given to appointing the oldest son from the second hut – but before this could happen Mangope appointed Edward 'Bob' Mogale, who just happened to be his uncle.[16]

Details of the lease with Lonmin are not for public disclosure, but the approximate amount paid in royalties to the baPo up to 2011 totalled R500 million, and the company currently pays the baPo R40 million per annum which is still held in trust by the North West provincial government.[17] This practice has been further entrenched in the Traditional Leadership and Governance Act which empowers provincial administrations to act in cases where there is evidence of maladministration by traditional authorities. BaPo potential mineral resources, to which Lonmin, International Ferro Metals and Samancor Chrome own mining or prospecting rights, are estimated at R10 billion. This makes them potentially one of the richest traditional authorities in the province.

By 2008 the political and financial affairs of the chiefdom were in disarray. A faction in the royal family opposed Edward, and all attempts to form a functioning traditional council failed. The provincial government appointed an administrator (again, something it was empowered to do through provisions in the Act of 2005). Before he left office, the administrator appointed a financial advisor to take charge of baPo financial affairs. It was to prove an ill-advised step. In October 2009, the appointee, Makepe Jeremiah Kenoshi, managed to change his status by concluding an agreement with the baPo traditional authority in which he was appointed as chief executive officer of the authority – apparently he was able to pull off this 'coup' through the close relationship he had struck up with Bob Mogale and others of his inner circle.

Relations between the baPo traditional authority and Kenoshi deteriorated quickly. He was accused by his employers of exceeding his authority and simply ignoring the baPo traditional authority, and despite their best efforts to block his financial dealings he signed an agreement with a company called Henti 2784 which planned property development in the region of the Hartebeespoort Dam, west of Pretoria. The amount involved was said to be in the region of R326 million, of which the baPo had to pay over R234 million by April 2010.[18] Kenoshi thus became both referee and player in baPo affairs and the baPo traditional authority had (probably good) grounds for believing that Edward would also benefit from Kenoshi's dealings. The 'royals' were pitted against senior members of their own community, represented by the baPo traditional authority, and many of whom were actually relatives of the chief.

The baPo traditional authority instituted disciplinary hearings against Kenoshi, at which point the provincial government stepped in and demanded that they be suspended. The North West premier, Maureen Modiselle, justified the action on the grounds that the province had to 'ensure proper administration by a traditional council' and that it was not her intention to 'loot the wealth of the traditional community for [her] self-enrichment'.[19] This led to a legal challenge by the baPo traditional authority to establish exactly who had *locus standi*. The North West High Court ruled against the baPo traditional authority and found in favour of the *kgosi*. Undeterred, the baPo traditional authority, through its attorneys, then issued an interdict against Kenoshi which was heard in the Gauteng High Court. The lawyers for the baPo traditional authority argued that Edward 'did not possess the mental capacity to discharge his duties' and 'had no control over the community's cash assets' said to be in the region of R300 million.[20] The court formed a different interpretation. The judge found that the baPo traditional authority *did* have the authority and status to institute proceedings against Kenoshi, and that the intervention of the provincial government was 'motivated by panic'. He argued that Edward was under the influence of people who might gain access to the R300 million fund, and that the fears of the baPo traditional authority were justified.[21] Kenoshi was subsequently fired by the baPo traditional authority.

Although *Kgosi* Edward was reconciled with the baPo traditional authority this was by no means the end of the matter. The provincial government, perhaps irked by the reverse suffered in the Gauteng High Court, imposed yet another administrator, Thabo Lerefolo, upon the baPo. Lerefolo was tasked with 'resuscitating and improving the administrative systems of the traditional council and the community, [and] managing and overseeing all the commercial activities, including the mining interests, of the community'.[22] From the perspective of the baPo traditional authority little had changed and they continue to have limited control over their assets, about which there is ever-growing resentment.[23] Lerefolo's term expired in October 2011, and a third administrator was appointed to run the affairs of the baPo. According to Bob Mogale, the administrators appointed by the North West provincial government have managed the chiefdom's business interests without consultation with the baPo traditional authority.[24]

In February 2012, Hugh Eiser, the lawyer for one of the baPo factions, aptly summed things up by saying that 'the premier [of the North West Province], the [members of the executive council] and the North West officials have done as they pleased with the community's money, while the community, which owns the money, is kept in the dark'.[25] Eiser then approached the public protector, Thuli Madonsela, asking for her intervention in securing the release of the trust funds held by the provincial government. *Kgosi* Bob, meanwhile, moved to the more comfortable surrounds of the Pecanwood Estate, near the Hartebeespoort Dam, in 2011, and has played little part in baPo affairs.

Madonsela began her forensic audit into the financial affairs of the baPo in mid-2013. The preliminary findings were startling, to say the least. It appeared that the various mining houses had deposited R400 million into the D account over the preceding four or five years, much of which had been siphoned off. Some R80 million had allegedly been used to build a 'royal palace' for the *kgosi*. The role of the North West provincial government was placed under scrutiny, particularly as it transpired that the by now notorious D account had never been audited and because Madonsela accused officials of refusing access to relevant documents. The North West premier, Thandi Modise, went on the defensive, stating that the North West provincial government was in the process of auditing the account itself, and accusing the public protector of wild 'allegations' regarding the missing funds. In late 2013 the matter was still unresolved.[26]

In another development, Eiser and the faction of the baPo that he represented in 2013 announced plans to take Lonmin to court to recover more than R100 million they argued was owed to the baPo for royalties which were avoided or not paid. This resulted from alleged underhand collusion between Lonmin's two subsidiaries, Western and Eastern Platinum Ltd. Through a complex rearrangement of sales procedures, Eastern Platinum sold its mined ore to Western Platinum at below market-related prices, leading to a loss of 8 to 10 per cent of its income and a concomitant drop in royalties payable to the baPo. The baPo do not receive royalties on mining by Western Platinum.[27] The company vehemently denies such a charge and countered that, if anything, there had been an overpayment of royalties to the baPo.

FIGURE 20: Lonmin Processing Plant, Marikana

Source: The authors

The baPo have also objected to the fact that Lonmin was not employing locals at its Marikana mine, an accusation that the company rejected. A structure called BaPo ba Mogale Unemployment Forum held a series of demonstrations and disruptions locally and nationally, which forced Lonmin into negotiations with the Forum to see how recruitment levels from the baPo could be increased.[28] Under some pressure from investors, Lonmin announced at the African Mining Indaba held in Cape Town in February 2011 that the company had created 643 jobs for the 'residents of communities around its operations' in the North West Province at the end of 2011.[29]

Finally, we see in the baPo case a potentially destabilising trend regarding the intrusion of broad based black economic (BBBE) interests into the mining sector. Lonmin established Incwala in 2004 to conform to the Minerals and Petroleum Resources Development Act which required mining companies to establish BEE structures. The baPo were given a 2.8 per cent share in Incwala in 2004. When the price of platinum fell in 2008, three of the other BEE partners sold their shares. At this point Mirabol, an entity established by Lonmin whose directors are appointed by Lonmin and which is associated with a trust also established by Lonmin, should have exercised whatever pre-emptive rights the baPo possessed when other shareholders sold their shares. The community did not have a direct say in what either the trustees or the directors of Mirabol did or did not do. Lonmin then proceeded to lend Cyril Ramaphosa's Shanduka[30] R2.5 billion which enabled it to acquire a 50.8 per cent share in Incwala in 2010. Shanduka contributed R300 million from its own coffers to clinch the deal. Thus the baPo failed to exercise their pre-emptive rights to purchase Incwala shares – which might have been avoided if Lonmin had approached the community directly.[31]

In addition to all these misfortunes the baPo had to endure the settlement of thousands of miners and residents of informal settlements on their land – one of the most vexed issues facing the baPo community as a whole. Prior to April 1994, outsiders who came on baPo land were male miners and they lived in mine hostels; non-miners from outside who wanted to settle on baPo land had to seek permission from the baPo traditional authorities and were allocated a stand in one of the baPo villages. After 1994 there was a huge influx of immigrants into the mining areas and the baPo authorities were powerless to prevent them from illegally settling on their land. In the estimation of Ezekiel Maimane, a royal family member, the population of baPo people in Wonderkop alone is not more than 5 000 whereas the immigrants number approximately 35 000.[32] These 'settlers' are from all over southern Africa, but especially the Eastern Cape, Mozambique, Lesotho and Swaziland. According to Maimane: 'We don't have a good relationship with these people'. The problem was worsened because many miners left their hostels in order to be eligible for the special allowance the Lonmin management gave to those miners who resided outside the hostels. In fact, even before the introduction of the allowance, many miners chose to live in shacks in the squatter settlement as well as their rooms in the hostel – a Bench Marks Foundation Report conducted in 2007 concluded that 'workers take the bare minimum of the allowance, stay in shacks and now have more money, either to send home as remittance or simply for local entertainment in the form of women and alcohol'.[33] The report also confirmed the high incidence of crime, including rape, and communicable diseases like TB and HIV/AIDS which have afflicted the general community as a result of the rapid increase in numbers.

The baKubung ba Monnakgotla: Unresolved divisions

The baKubung ba Ratheo based at Ledig near the well-known resort of Sun City have also been uncertain beneficiaries of platinum mining. The baKubung ba Ratheo were removed from their farms at Molotestad near Ventersdorp in the former western Transvaal in 1967.[34] In exchange for the loss of their farms they were given alternative land, comprising the farms Ledig 909 JQ (where the majority actually settled), Koedoesfontein 94 JQ, and Wydhoek 57JQ, although there was some confusion initially as to whether they were entitled to settle on the latter two properties. Although one faction of the community refused to move at first the entire population had been resettled by 1969.

The affairs of the baKubung were officially handled by their *kgosi*, David Gabonewe, until he died in 2009. In 2003, however, he had suffered a stroke and after consultation with his six daughters and other senior members of the ruling family, he chose his second oldest son, Ezekiel, to act until such time as a more substantive appointment was made. David had an older son, Solomon, but both his father and the majority of his followers, it seems, favoured Ezekiel. Solomon's unpopularity related to his personal unsuitability for the position, and to the fact that he had been identified with the disagreeable (in most people's eyes) regime of the former homeland president Lucas Mangope.

During Ezekiel's tenure the fortunes of the baKubung rose dramatically. Until this time they had been an impoverished community, subject to the dubious benefits of being so close to the glitz of the Sun City resort. Then prospecting on their farms revealed mineral potential. In 2004, the community was given a 33 per cent stake, or 117 million shares amounting to R500 million, in Wesizwe Platinum, a company listed on the Johannesburg Stock Exchange, thus catapulting the baKubung into the heady world of boardroom financial dealings and politics. When David Gabonewe died in 2009, the struggle for the community's assets intensified. Subsequent developments regarding the assets of the baKubung led to investigations that revealed a Byzantine state of financial cover-ups and subterfuge. In what the *Citizen* newspaper of November 2010 called the 'year's most under-reported domestic scandal' it was revealed that in 2007 the baKubung traditional council headed by Ezekiel had appointed a firm called Musa Capital, a company owned by two US citizens, Antoine Johnson and William Jimerson, to act as financial advisers to the baKubung. Musa then 'monetised' the shares held by Wesizwe allocated to them under the conditions of the original BEE deal. The transaction was effected ostensibly to diversify the baKubung's asset base. About 44 million shares, at R8 a share, were then used as collateral to obtain a loan from the Deutsche Bank for approximately R400 million. A second transfer of shares occurred in March 2009. Using the same process, Musa obtained R296 million for the balance of the shares held with Wesizwe – this time, however, the cash loan was provided by the Industrial Development Corporation of South Africa. Musa created an array of financial entities: New Shelf 925, TransAfrica Mining Ventures, Hippo Mining Ventures, to name but a few, operating in countries as far distant as Argentina, to facilitate the dissipation of the nearly R700 million realised by the conversion of the shares held by Wesizwe into cash.[35]

Musa Capital drew a veil of secrecy over the location of the cash realised by the deal, claiming that it was subject to a number of 'confidentiality' agreements between it and the traditional council, and that to reveal the source of the

FIGURE 21: Bakubung Wesiswe Platinum Mine near Ledig, with Sun City and the Pilanesberg range in the distance.

Source: The authors

investments would be tantamount to revealing information to competitors. This invited the attention of the North West provincial government. In March 2010, Premier Maureen Modiselle appointed an administrator to investigate any unauthorised expenditure and/or withdrawal of money belonging to the baKubung. When Musa still refused to cooperate with the investigation, the matter was referred to the High Court in Mafikeng, which ordered Musa to answer the questions asked of it by the administrator, and to furnish it and a group opposed to the traditional council (known as the Bakubung Concerned Group) all documentation regarding the 'monetisation' of the shares. Still unhappy with the content of the documents provided after the ruling, the Concerned Group was able to force the matter into an arbitration ruling. Apart from the 'disappearance' of the cash held under Musa's control, the administrator's report revealed other disturbing details about the way in which the financial affairs of the baKubung had been conducted; in particular that about R25 million in donations from Wesizwe meant for development projects could not be accounted for.[36] According to the lawyer acting for the traditional council, Lucas Moalusi, while there may have been undue influence placed by Musa on Ezekiel Monnakgotla, there was no evidence of corruption in any of these dealings. Moreover the loans given by the Deutsche Bank and the Industrial Development Corporation were beneficial to the baKubung, as these institutions bore the risk if share prices were to decline.[37]

Caught in the crossfire of the fallout, Wesizwe was riven by a bitter contest for boardroom control. The two baKubung appointees to the Wesizwe board, *Kgosi* Ezekiel Monnakgotla and an ANC politician and businessman, DJ Phologane, influenced the board to remove the chief executive officer, Mike Solomons, and chairperson Robert Rainey. The two had accused Musa Capital of engineering a 'hostile takeover' of Wesizwe, and were clearly unhappy with the way the shares had been 'monetised' by Musa. The board then appointed two new incumbents, economist Iraj Abedian and Nyasha Tengawarima. The removal of Solomons and Rainey was challenged by Kubung opponents of the traditional council, who claimed Ezekiel Monnakgotla and Phologane were not the elected baKubung representatives on the board of Wesizwe. The battle for control is still undecided. In addition, however, the Deutsche Bank and the Industrial Development Corporation were accused of inadequate financial oversight for the granting of the loans. Suspicions that other political and business connections were also involved in the entire deal were raised by revelations that the wife of the one-time minister of social development, Zola Skweyiya, and the president of the Chamber of Mines, Sipho Nkosi, had substantial stakes in companies linked to Musa Capital.[36] Moneyweb, through its international Internet site, Mineweb (which exposed much of the content related to the unusual activities of Musa) was accused by the company of failing to ascertain the 'facts' behind the case and for lack of objectivity in its reporting. Musa's lawyers insisted that 'none of the [baKubung] community's money has disappeared', and demanded a retraction of the allegations made by Moneyweb and Mineweb.

It becomes apparent, therefore, that the disposal of the baKubung assets tore an already fractious community into new opposing factions that pitted family members against one another. The Concerned Group accused Ezekiel of 'secrecy and dismal mismanagement of the community assets'[39] because of allegations that a select few of Ezekiel's inner circle and financial confidantes were enriching themselves at the expense of the rest of the baKubung ba Rantheo. Other lesser charges, mainly to do with his disruptive personal conduct, were levelled at Ezekiel. The dispute, on one level at least, was one between different components of the 'royal' family, and who was entitled to represent the baKubung. Ezekiel's accusers found a figurehead in Margaret Monnakgotla, his sister, who was backed by a few of his uncles. Ezekiel, on the other hand, had the support of all the sisters of David Gabonewe and a couple of his brothers. To an extent, the feud was also one between the defenders of the traditional order in the person of Ezekiel – who controlled the traditional council – and a modernist faction, largely urban, headed by a distant uncle, Mokgwari Monnakgotla, resident in Johannesburg, who had significant support in the *khuduthamaga* or royal council. Ezekiel's backers countered that the Concerned Group had little interest in the affairs of the baKubung until the mining deals were struck, and that their sudden interest was prompted by the prospect of employment and other opportunities.

These disputes culminated, on 29 January 2010, with the acting *kgosi's* receipt of a letter from people purporting to be members of the royal family (the Concerned Group) indicating their decision to remove him from his position and to replace him with his sister Margaret. This decision was communicated to the premier of the province who agreed and withdrew Ezekiel's certificate of recognition. Ezekiel and the traditional council then urgently approached the North West High Court in an attempt to prevent the premier from publishing a notice of her withdrawal of recognition. They were unsuccessful, but in mid-2010 Ezekiel instructed his lawyers to challenge the High Court ruling in the Supreme Court. The culmination of all this wrangling was that the North West High Court in Mafikeng ruled that the traditional council should account for the location of all the 'missing millions', and an administrator was appointed to oversee the process. After the submission of forensic reports, the administrator recommended that legal action be taken against Musa and the traditional council.

The two matters (the attempt to force Musa to reveal the location of Kubung assets, and the attempt to oust Ezekiel) were thus linked, and unresolved in 2010, and the community was very divided. Crucial to determining the validity of the decision was the structure of the baKubung royal family, for Ezekiel's supporters claimed that the *khuduthamaga* was packed with distant and non-core members when the decision to remove him was taken, and that the correct procedures were not followed. The legal issue thus hinged on views regarding the interpretation of what constituted a royal family as set out by the Traditional Leadership and Governance Framework Act 41 of 2003, and the right of the *khuduthamaga* to confer chieftainship on a woman.

In May 2012 Ezekiel's position became untenable and the premier of the province, Thandi Modise, withdrew his certificate of recognition as chief. This time even his staunchest supporters deserted him – but nobody has been appointed in his place. The former backers of Ezekiel switched allegiance to his brother Solomon, who is genealogically the rightful heir. He is, however, viewed by the Concerned Group as being under the influence of his brother Ezekiel and Phologane; more significantly, he is regarded as unfit for chieftainship in the sense that he does not possess the requisite business skills or educational background to rule over the kind of mineral-rich community that the baKubung ba Ratheo has become. While the provincial government dithers in its attempts to identify the Kubung's rightful representatives, they remain rudderless. It remains to be seen who will control the considerable financial reserves of the baKubung ba Ratheo.

The baKgatla ba Kgafela: 'A recipe for another Marikana'?

The development of mining and, to a lesser extent, tourism-driven revenues in the region has dominated the economic and political developments of the baKgatla as well. In the mid-1990s, the baKgatla negotiated with Anglo Platinum for the payment of R6.2 million per annum for royalties on mining operations on farms they owned. In addition they derived a fairly important income stream from revenues earned from controlling an entry point, the Bakgatla Gate, into the bordering Pilanesberg National Park. This resulted in the accumulation of over R60 million in the community coffers by 2007. In 2005, the chief executive officer of Anglo Platinum (Angloplats) announced that the baKgatla had acquired a 15 per cent equity stake in an Angloplats operating asset with effect from December 2006.

A more significant development emerged when Platmin was created in 2009 with a primary listing on the Toronto Stock Exchange and secondary listings on the London Stock Exchange and the Johannesburg Stock Exchange. As soon as Platmin was created, the baKgatla ba Kgafela traditional authority invested R500 million in it through a joint venture with Palingust Investment Consortium, which has a stake in Platmin. With a 32.98 per cent stake in Platmin, the baKgatla ba Kgafela traditional authority quickly became the consortium's single largest investor. Subsequently, the baKgatla stake in Platmin has risen to 50.1 per cent and is owned through the Moepi Group, in which the baKgatla community is the major shareholder. The Moepi Group is a mining investment vehicle that owns 27.6 per cent of Boynton Investments, the South African operating structure of Platmin through which the South African-located assets are all owned.[40] In May 2011, *Kgosi* Nyalala Pilane was appointed to the board of directors of Platmin,[41] an important development which, at least symbolically, indicated some form of baKgatla say and control over mining operations on their properties. Prospects for the baKgatla, or at least certain designated factions of the chiefdom, looked rosy.

In order to deal with the unfolding economic and financial opportunities provided by platinum mining on their lands, the baKgatla leadership reorganised themselves into a corporate entity run along business lines, following in the footsteps of their neighbours, the baFokeng. In 2008 they created the baKgatla ba Kgafela Traditional Authority, the corporate business entity. Its objective, as outlined in its 2010 strategic plan, was to bring about socioeconomic advancement and to improve the quality of life of its community through mining, agriculture, industrial development and tourism. It also aimed to promote the participation of local entrepreneurs in industrial development and service provision; assist in the provision of skills to the baKgatla to enable them to utilise economic trends and opportunities; create an effective governance system and infrastructure for improved coordination and communication amongst the various Kgatla villages; and develop better ways of investing resources.

The assets accruing from mining on Kgatla land were used in several ways to improve the lives of the villagers. They included the construction of a stadium seating 30 000 people at a cost of R174 million, administration offices, a bulk sewage reticulation plant, and the construction of roads. A number of clinics were built or revamped in Kgatla villages. In addition to this, the mines have made significant contributions to the overall formal education of baKgatla, principally in the provision of equipment or in improving the qualifications of educators.[43] The baKgatla ba Kgafela Traditional Authority also funds the education of baKgatla students at tertiary institutions, and in 2009 launched a mining academy run jointly with the nearby Mankwe Campus of Further Education and Training in Mogwase. Many more development initiatives are planned under an ambitious project called the Greater Moruleng Revitalisation Project.[44] In many respects, then, the baKgatla are moving towards the model of the corporate traditional community espoused by the baFokeng.

However, there emerged tensions and disputes that mirrored those that afflicted the other major black beneficiaries of mining in the Rustenburg bushveld mineral-rich complex. In May 2006, the baKgatla *kgosi*, Nyalala Pilane, was charged, along with his cousin and two co-accused, with forty-two counts of theft, corruption and fraud. It was alleged that they had defrauded the baKgatla of over R40 million. The Kgatla mining consortium had approached the South African Land Bank for a loan, estimated to be over R40 million, with which they intended to purchase a number of farms. Pilane allegedly undertook to repay the loans from the Land Bank when he received the annual net income from royalties of R6.2 million, but according to certain members of the 'royal family' this had never met with the approval of the community. On the basis of this complaint, and a demand from the Land Bank that Pilane repay the debt, the premier of the North-West Province, Edna Molewa, appointed an auditing firm to investigate. They referred the matter to the police and recommended a prosecution. It also emerged from the audit that Pilane had, among other alleged offences, undertaken a trip to Australia paid for by the premier's office in Mafikeng that was not authorised by the baKgatla council. Another withdrawal of R70 000 from the baKgatla account, although approved, was allegedly misspent by Pilane. It was reported by the South African Broadcasting Corporation in a news bulletin on 26 May 2006, that the Asset Forfeiture Unit of the Scorpions had attached Pilane's assets.[43] The case was costly to Pilane and his co-accused, largely because it dragged on until mid-2008. In the interim, representatives of the baKgatla sent a memorandum to Edna Molewa requesting Pilane's removal from office – the provincial government declined to act, however, until the trial had run its course.[45] Feelings among the baKgatla were inflamed and the community was divided. At Sun City, a meeting addressed by Pilane was disrupted by a crowd of enraged Kgatla villagers, and sixteen people were arrested for public violence.

FIGURE 22: Nyalala Pilane, current *kgosi* of the
baKgatla ba Kgafela.
Source: Mphebatho Museum, Moruleng

In February 2007, Kobedi Pilane was acquitted of theft and fraud charges but
in June 2008 he was found guilty of theft of tribal funds and of defrauding the
Land Bank. He took the matter on appeal and won his case. It was not in any
sense a real exoneration, for the court found 'much criticism against the way in
which [Pilane], as the person in charge of the tribe's money, administered it'.[47]
The charge of corruption against him was withdrawn and he continued to protest
his innocence.

Another setback to Kgatla fortunes was to follow. The deal between *Kgosi*
Pilane and Anglo Platinum was derailed by a legal challenge from a smaller off-
shoot of the baKgatla ba Kgafela, the baKgatla ba Sefikile, numbering 1 400 people,
who claimed to be the descendants of the inhabitants of the original farm, Spitzkop, on which mining operations were scheduled to commence. Just as several
independent groups among the baFokeng were later to challenge ownership of
farms claimed by the Fokeng ruling family, so the baSefilike claimed that 'fifty-
two members of this tribe [the baSefikile] collected money and purchased land in

1911, which was registered in the name of the chief in trust for the tribe'.[48] Their lawyer, Richard Spoor, claimed that the baSefikile had no knowledge of the lease agreement between Pilane and Anglo Platinum, who would not make the terms of the lease available to his clients. As was to be the case with the baFokeng, the ruling house of the baKgatla and Anglo Platinum denied that 'any such entity or group exists'.[49]

The matter flared up once again towards the end of 2010, when further information came to light and shifted the issue onto a graver and potentially more damaging and disputatious level. It emerged that Anglo Platinum had in fact been mining on Spitzkop illegally, something which the company became aware of in 2006 but apparently never disclosed. Initially, an attempt was made to negotiate compensation for the encroachment, but finally Anglo Platinum agreed to lend the baKgatla R45 million, three of their farms being put up as surety for the loan. The R45 million was allegedly not deposited into the bank account of the baKgatla but into the trust account of their attorneys.

These transactions appeared also not to have been disclosed to most members of the chiefdom until 2010, although Nyalala Pilane and Anglo Platinum denied it and countered that all transactions between the two parties were approved by a general meeting of the entire baKgatla *morafe* in November 2006. Significantly, the controversy now extended to include many other members of the baKgatla and not just the baSefikile. It closely resembled developments among the baKubung; the alternative 'royal family' came increasingly to the fore, headed by two other member of the Pilane family, *dikgosi* Phetoe Pilane and Thari Pilane, who demanded to see the books of the Bakgatla ba Kgafela traditional authority. This alternative royal family questioned Nyalala's legitimacy, claiming that his appointment as *kgosi* of the baKgatla ba Kgafela had expired and that he should resign to make way for a new incumbent.[50] The vexed issue of the legitimacy of the Moruleng chiefship has not been resolved despite attempts by Nyalala Pilane to bring the feuding royal houses together. Phetoe Pilane has gone so far as to repudiate the authority of the original baKgatla based in Mochudi, Botswana, who confirmed Nyalala's appointment, claiming that they should not have power over the affairs of the baKgatla in another country.

In a not unrelated development, another faction of the baKgatla also refuted the authority of Nyalala Pilane. In 2010, the baKgatla ba Kautlwale attempted to constitute itself as a separate and independent ethnic entity from the Bakgatla ba Kgafela, basing this claim on the argument that they had been independent prior to the imposition of 'grand apartheid', after which all baKgatla factions were wrongly incorporated into the baKgafala,[51] and calling themselves the Motlhabe Tribal Authority to reflect their independence. However, when they attempted to convene a community meeting to discuss the possibility of secession from the baKgatla ba Kgafela, Nyalala Pilane allegedly prevented it from taking place by calling in the police. According to the baKautlwale, Nyalala Pilane had also refused to attend meetings to discuss the issue with officials from the North West provincial government, and continued to exercise his authority in the baKautlwale villages through his own appointed *kgosana* (headman) whose legitimacy was not accepted by the local people. Pilane subsequently obtained an interdict preventing the baKautlwale from holding traditional meetings, claiming that the baKautlwale affirmation of independence was a fiction; indeed there is little in the official records to confirm their existence as a discrete community.

Towards the end of 2010, the baKautlwale approached the Legal Resources Centre to advance their case for independent status. A researcher for the Legal Resources Centre, Aninka Claassens, identified the Pilane case as an acid test for the whole issue of 'unaccountable chiefs' in the platinum belt, asserting that they were 'a recipe for a new Marikana'.[52] The matter was heard in the Constitutional Court towards the end of 2012. In a landmark ruling, the court (with two justices dissenting, including the chief justice) set aside the interdicts Pilane had obtained, thus affirming the right of the baKautlwale to secede if they so wished. Justice Thembile Skweyiya, however, urged the parties to resolve their differences through dialogue.[53]

Behind the baKautlwale's desire for a degree of autonomy lie deeper material interests. The farm on which many of them live, Witkleifontein, is known to be platinum bearing. However, it is subject to a land claim under the land restitution policy lodged by Nyalala Pilane in the name of the baKgatla ba Kgafela which, the baKautlwale argue, has excluded them. In court papers lodged in 2011, the baKautlwale contend that they have an interest not only in the land claim for Witkleifontein but also in respect of two other farms, Rhenosterkraal and Welwewaght, which they assert were purchased with contributions from their former *kgosi* Kautlwale and a number of his followers in 1911 and 1926 respectively. In conclusion, the baKautlwale declare that 'baKgatla interests in platinum are on properties where we live … but despite this our village is extremely poor and undeveloped … we do not see any of this money and how it is used'.[54] Another complaint they lodged was that proceeds from infrastructure that they had developed on their land, including three schools, a clinic and a community hall, had all gone into the coffers of Pilane's baKgatla at Saulspoort. Yet again, the case mirrors that of the challenge to the Royal Bafokeng Administration's attempts to register trust farms and transfer land in the Phokeng region into its name. The windfall from mining, for the black inhabitants of the wider Rustenburg region, has provided both the motive and the means for incumbent *dikgosi* to tighten their grip over these resources, in turn forcing a political challenge from opponents who for varying reasons feel excluded from the profits, or alternatively wish to gain a larger share for themselves.[55]

Finally, there is an added dimension to ethnic tensions and conflict over the benefits of mining among the baKgatla that gives a unique twist to the story. The seat of Kgatla authority lies in Botswana. The paramount chief in Mochidi, Kgafela Kgafela, has viewed mining revenues as being jointly owned by both sections of the baKgatla ba Kgafela, in Botswana and South Africa. A clear indication of the extent of Kgafela's chiefly jurisdiction over the Pilanesberg baKgatla is the fact that in 2012 he took the baKgatla ba Kgafela traditional authority's attorney to the North Gauteng High Court to compel him to disclose 'how he received R49 million from the baKgatla tribe' as shown in a recent financial audit. Furthermore some of the funds held by the baKgatla ba Kgafela traditional authority were given to members of the royals in Mochidi to pay for legal fees regarding court cases they were facing in Botswana.[56]

Then, towards the end of 2012, Kgafela Kgafela fled Botswana for Moruleng after a warrant of arrest was issued for his failure to attend court in a case involving his alleged assault of some villagers in Mochudi. Initially, the two *dikgosi* exuded an image of unity, but tensions arose when Kgafela 'uncovered abuse claims against his close friend', Pilane. His interference in the internal affairs of the Pilanesberg baKgatla was resented by Pilane (who by this time had raised his profile by being elected deputy chairperson of the ANC-aligned Contralesa, the Congress of South African Traditional Leaders) and most members of the chiefdom in South Africa, and his actions have created bad blood between the royal relatives, to the extent that Nyalala Pilane regards the Pilanesberg baKgatla as being practically independent from the main house. Most baKgatla living in South Africa resent the fact that resources are being expatriated to the Botswana section of the chiefdom, and many are calling for a complete break between the two sections. At the end of 2012, because of the sensitive and deep-seated nature of Moruleng-Mochudi relations, the Moruleng leadership decided to leave the matter in abeyance but with the understanding that any final decision would be made by the community '[after] consultations with relevant stakeholders and [the] broader baKgatla ba Kgafela traditional community across the thirty-two villages'.[57]

These cases elaborated upon here are not the only instances of the troubled relationship between mining companies and traditional communities and the conflicts caused by distributing profits from mining to these beneficiaries.[58] Sharing the revenues from mining with traditional communities has not been very successful. The main problems revolve around misappropriation (or allegations of misappropriation) of funds; the composition of these *merafe* and their rightful leaders; and questions surrounding historical ownership of the land on which mining is taking place. These have been accompanied by numerous costly legal wrangles which have often delayed mining operations and the formation of partnerships that are ultimately to the benefit of the communities.

National parks and game reserves

Two significantly large national parks lie in the bushveld region, the Pilanesberg and the Madikwe. The Pilanesberg was the first to be established. We have mentioned in the preceding chapter how the park's establishment created tension between Mangope and the baKgatla after several families had to be removed from the Pilanesberg crater to make way for the scheme. The park had its origin in a report drawn up in 1969 by academics from the former Potchefstroom University and provided a material and ideological basis for legitimising what was to become the future homeland of the Tswana, thus 'easing [Bophuthatswana] into a place of respectability in the international community'.[59] The idea of the park became a reality with the construction of the Sun City resort in 1978, for the adjacent park provided a further attraction for visitors. In fact, the choice of the Pilanesberg volcanic crater was not a particularly appropriate one, for 'little of the natural flora and fauna [had] survived the wasteful pastoral practices of both whites and Africans',[60] and the park had to be almost entirely restocked. Nevertheless, through the efforts of capable and enthusiastic rangers, the Pilanesberg became a model for wildlife management, and generated jobs and a considerable income for the state.

Some compensation was given to those who had been removed to make way for the park. The 104 *oorlamse* families who had lived for over a century at Welgeval within the crater were given R230 000 for the loss of their church,

FIGURE 22: View of Pilanesberg crater from the north east.

Source: The authors

school and houses, and an additional R153 000 for the farm itself, an amount that was considered insufficient by the community. Some sixty families were moved to the farm Zandfontein, about ten kilometres east of Saulspoort, and the remainder to Rhenosterkraal and Cyferkuil, about thirty kilometres from Zandfontein, but they were unable effectively to resume agricultural activities and many lost cattle, either through lack of proper fencing or because cattle were rustled in the process of resettlement. Water resources on the new farms were also far inferior to those they enjoyed at Welgeval. The removal spelled the end for the independence of this close-knit and unique community.[61]

In a report commissioned by the park's management in 1984, the researcher concluded there had been little material benefit for the surrounding residents. This was despite the founding philosophy that the park should be integrated into the surrounding region and that it benefited its inhabitants. The conclusion was reached that 'some 50 000 people, or some 7 000 households, have been deprived materially by the creation of the park. On the other hand, there are the educational benefits of the park, of questionable material benefit, and the creation of about 200 jobs.'[62] This rather bleak outlook has been mitigated by the implementation of a more aggressive social responsibility programme begun in the late 1980s. In addition, the local community began to benefit from the culling programme; hunting fees were paid out to local councils and, in the case of the baKgatla ba Kgafela, they profited from entrance and camping fees from the one entry point they controlled.[63]

The Madikwe Game Reserve emerged in a rather interesting manner. In 1986 the Molatedi Dam was constructed about 100 kilometres north of Zeerust but, mainly because of its isolation from the markets, its potential was never realised by the surrounding farmers, whether black small landholders or white commercial producers (the farms of many of the surrounding white farmers had been expropriated to form what became known as the Marico corridor, an attempt to consolidate the impossibly fragmented homeland of Bophuthatswana). In a feasibility study conducted by a consultancy known as the Settlement Planning Services it was recommended that optimal use of the dam could be achieved by the development of a 'major wildlife-based eco-tourism destination which incorporated the "big five"– lion, elephant, rhino, buffalo, and leopard'.[64] Part of the rationale was that the game reserve would generate at least 1 200 jobs and inject R7 million into the local economy.

The reserve, which was constructed in 1991, lies about eighty kilometres north of Zeerust and abuts Botswana. As this was state-owned land, it did not involve the removal of people from their homes as happened in the Pilanesberg, although several communities subsequently claimed it as ancestral land. It comprised 75 000 hectares which had previously been what was described as 'degraded cattle farms'.[65] The local community comprised the villages of Lekgop-hung, Supingstad and Molatedi. Livestock was the main means of subsistence, but a survey revealed that 50 per cent of households owned no stock at all, and it was confined to a small minority who owned about a hundred head of cattle each.[66] The Madikwe reserve was run by the Bophuthatswana National Parks, now the North West Parks and Tourism Board.

From its inception the Madikwe Reserve presented itself as 'world-renowned for its people-based approach to conservation', in the belief that 'local communities and individuals must benefit significantly from wildlife conservation and related activities'.[67] The reserve was thus formed on the basis that it was 'three-way partnership between the state, local communities and the private sector'. The private sector would manage and develop tourism initiatives and activities in the reserve, while local communities were to be financially assisted, from the funds generated by the entity, to establish community-based projects which were meant to act as a 'major social and economic core … around which the entire development of the region can be based.'[68] This approach is part of a people-based conservation philosophy that has become the guiding principle for conservation in Africa generally.

The reserve has created between 500 and 600 jobs, five times more for the local community than that provided previously by the agricultural sector, but there have been problems, as the parks management is quite prepared to admit. Community involvement and development has been slower than anticipated through lack of local government structures. Initially, the local population had no defined property rights within the park itself, and thus could derive no financial benefit from conservation or tourism agencies. The expansion of the park was affected by land claims under the Land Claims Restitution Act, which placed the development of lodges within the park on hold until the land claims were resolved around 2000.[69] The three-way partnership was not in reality on a basis of equal status, as the private sector and the North West Parks and Tourism Board had far more clout than the local community – and of the positions created by the construction of private lodges not all were filled by the local residents, as most of the workers lacked the requisite skills. In order to solve these problems, the Madikwe Initiative, funded by the UK Department for International Development (DfID), was designed. Its first aim was to grant certain commercial rights within the park to the local communities, in order to allow them participation in the eco-tourism driven economy, and a second purpose was to give a collective and participatory entity to the communities in the form of a community development organisation which would administer the benefits derived from the park in the best interests of all the local inhabitants. Unfortunately, the community development organisation had to try and operate in an inflamed and conflict-ridden environment.

There were political tensions between social and political components within the villages (supporters of the new ANC government and the former Bophuthatswana regime) and between the villages themselves (over claims as to who was politically paramount). In addition, the communities felt that the development projects were top-down. This disillusionment was countered by plans to involve the local residents in lodge enterprises in partnership with private sector companies, and by setting up and supporting small businesses to provide materials for the construction of the lodges and other buildings in the park. Both of these initiatives met with mixed results. But nevertheless, as the report by E Koch and PJ Massyn concludes, the lessons learned at Madikwe 'can be replicated in other integrated conservation and development projects'.[70]

In the field of conservation in South Africa Madikwe has pioneered many new approaches including the re-establishment of species of wildlife, in particular the successful relocation of elephant family groups and large adult bulls, and the introduction of predators such as spotted hyaena and wild dogs from wild and captive-bred populations. Its successes and reputation should ensure its commercial viability, which should translate into tangible benefits for the local villagers of this very impoverished region of the bushveld.

Sun City

'Bophuthatswana is far away/But we know it's in South Africa no matter what they say!'[71]

From the very time of its conception in 1978 Sun City was fraught with controversy. Exploiting the opportunity created by Bophuthatswana as a state 'independent' from South Africa at the height of efforts to isolate the country, Sol Kerzner, already an established development mogul in the entertainment industry, constructed a resort that might attract many of the world's top class entertainers. Simultaneously, Sun City could provide the hidden fruits of gambling, pornography and illicit sex denied to South Africans by their Calvinistic and authoritarian government. Sun City was located only a couple of hours' drive from the major cities of Johannesburg and Pretoria and the idea was that it would match the more famous of its kind and become the 'Las Vegas of Africa'. Construction near Ledig in 1979 contrasted starkly with the surrounding poverty in a rural district run down by a surplus human population and an overstocking of grazing animals. Sun City, which hosted a number of international artists in the 1980s, became the focal point for the cultural boycott of South Africa.

The 1982 Keenan Report on the Pilanesberg Park highlighted (and probably magnified) some of the effects of the establishment of Sun City on the surrounding inhabitants. By 1984, unemployment at nearby Ledig, although still high, was below average for the region, indicating that its proximity to Sun City did provide some employment for the residents – many of the more skilled workers, however, were drawn from the Reef and Pretoria.[70] Moreover, the local inhabitants considered wages at Sun City to have been lower than those offered elsewhere, in particular nearby Rustenburg, and argued that working conditions were worse. So dire were workers' circumstances, in fact, that in 1982 they selected a five-person delegation to express their grievances to the general manager, Peter Wagner. Despite promises to improve the situation, nothing was done, and the delegation approached Mangope.

The Bophuthatswana government's reaction was initially quite sympathetic, and the president sent his minister of manpower, Rowan Cronje, to investigate the situation at first hand. Cronje envisaged a solution through the creation of a workers' representative committee. One was formed, but much on the basis of one that had failed before owing to lack of legitimacy and support. A second attempt to gain Mangope's intercession failed totally – the delegation was accused of having been influenced by Dr Motlana's Soweto Committee of Ten, a political pressure group which had also gone to the press regarding working conditions at Sun City. The bad publicity annoyed Mangope, and Wagner fired the five members of the delegation, presumably with the blessing of the president. Such a response was further proof to the inhabitants of Ledig and other surrounding villages that Bophuthatswana was indifferent to their plight, much as it had been to the consequences of the formation of the Pilanesberg reserve.

To close this short discussion on the impact of the Sun City mega-resort, we should briefly mention that its establishment triggered a worldwide cultural boycott to prevent artistes from Europe and the United States from performing at the venue. This initiative was hampered by a shift to more conservative political administrations in Britain and the US and by Bophuthatswana's ability to successfully project itself as an independent state, with no relation to apartheid South Africa. While the cultural boycott generated considerable political heat and (generally adverse) publicity for Sun City, it does not appear to have fundamentally changed the relationship between the resort's management and the surrounding black population living in the nearby villages.

From the mid-1990s, and coinciding with the dawn of a democratic order, Sun City became more aware of the glaring inequalities between its wealthy visitors and the poverty of the surrounding inhabitants, and sought to reduce this discrepancy (and pre-empt acts of violence and theft) by uplifting the quality of their lives through an ambitious social responsibility programme. Sun City built and equipped three clinics nearby, as well as three hospices, able to cater for about thirty patients. The staffing of the facilities is the responsibility of the provincial Department of Health, and the Department of Public Works maintains the buildings. It also built an additional ten classrooms at the high school in Ledig, and constructed and equipped two early learning centres in the same village. A middle school was erected in Moqwassie to augment the existing Holy Family High School, and additions and refurbishments made to schools in Tweelaagte and Mabieskraal. A primary school was built at Mavela near Maqwassie, where in addition sixteen houses were built for child-headed households. The resort was also responsible for the construction of a police station at Ledig. These facilities are also run and maintained by the relevant provincial departments. Sun City staff furthermore train and advise the owners of bed and breakfast establishments in customer and financial services.[73]

This indicates a more responsive policy on the part of Sun City to the needs of the poorer residents in the vicinity, and presumably has served to moderate ill-feeling among them about the opulence of the resort. About three million people visit Sun City annually which represents a sizable injection of cash earnings into local tourist and service facilities. The resort has 1 400 permanent employees, a few drawn from the local community, and about 4 000 casual workers at peak periods, almost all residents of the surrounding villages.

ENDNOTES

1   A part of the first section of this discussion is drawn from 'Mining and Traditional Communities in South Africa's "Platinum Belt": Contestations Over Land, Leadership and Assets in North-West province, c. 1996-2012 *Journal of Southern African Studies*, 39, 2, (2013), pp. 409-423.

2   Cited in I. Kriel, '"Visible Through Ethnicity": Representations of the "Royal BaFokengNation"', Seminar Paper, Pretoria University, September 2008.

3   Cited in I. Kriel, op. cit., p. 5.

4   J.L. and J. Comaroff, *Ethnicity Inc.* (Chicago: University of Chicago Press, 2009), p.116.

5   S.E. Cook, 'The Business of Being BaFokeng: The Corporatisation of Tribal Authority in South Africa', *Current Anthropology* 52 (suppl 3) S151-S159 .

6   S.E. Cook, 'Chiefs, Kings, Corporatization and Democracy: A South African Case Study', *Brown Journal of World Affairs*, Vol. 12, No. 1 (2005), p. 136.

7   This is a principal argument advanced by Kriel in 'VisibleThrough Ethnicity'.

8   Royal BaFokeng Nation, Population and Use of Land Audit, (PULA), Study, 2011, p. 5.

9   Op. cit., p. 17.

10. S. Mnwaba, 'Participation and Paradoxes: Community Control of Mineral Wealth in South Africa's Royal baFokeng and baKgatla ba Kgafela Communities. (D. Phil Dissertation, University of Fort Hare, 2011).

11 The basic facts are set out in a report in the *Sunday Independent*, 10 November 2010, 'The Dispute that Could Re-write the History Books'.

12 Cited in Manson, 'Platinum Mining', p. 411. From http:/www.anc.org.za/anc-docs/history/mandeka/1995/sp951126c.html.

13 See Responding Affidavit of G.J. Capps, Case No 999/2008. Mafikeng High Court, In the Matter Between the Royal BaFokeng Nation and (Applicant) and the Minister of Rural Development and Land Reform, (1st Respondent).

14 Cook, 'The Business of Being BaFokeng', p. 16.

15 As a result of the baFokeng case regarding royalties the assets accruing to them were removed from the D account.

16 Interview with Mr Ratsheke Maimane, Wonderkop, 7 February 2013, with Andrew Manson and Bernard Mbenga.

17 Lonmin Press Release 21 August 2011 by Bernard Mokoena, Executive Vice-President, Human Capital and External Affairs. http://www.lonmin.com/down-loads/media_centre/news/press/2011/21aug11-baPo_ba_mogale.pdf.

18 See North Gauteng High Court, Pretoria, Regional Authority of the BaPo Ba Mogale v Kenoshi and the Premier, North West Province, Case no 31876-10 2010, July 2011. Paras 20-25.

19 South African Government Information, BaPo ba Mogale media briefing by M. Modiselle, 5 October, 2010. http://www.info.gov.za/speech/DynamicAction?pageid=461&sid=1341.

20 Case 31876-10, para 15.

21 Op. cit., para 19.

22 New Age Online, http://thenewage.co.za/16868-1017-53-Lawyer_takes_charge 2 may 2011.

23 Personal communication with H. Eiser, attorneys for the TA.

24 *City Press*, 12 February 2012, 'Madonsela to probe tribes' lost millions'.

25 Cited in *City Press*, 12 February 2012.

26 See *Sunday World*, 30 September 2013, 'Madonsela: Mining Royalties Dwindle from R400 to R16 m', htpp://www.enca.com/south-africa/madonsela-minig-royalties-dwindle.

27 *Mail & Guardian Online*. http://mg.co.za/article/2012-11-23-00-baPo-community-wants-lonmin-right. http://mg.co.za/article/2012-12-14-00-lonmin-faces-community-showdown.

28 See *Sowetan*, 28 August 2011, 'Lonmin Agrees to Employ More Locals'.

29 See *Business News*, Business Report, 13 February 2012, 'Lonmin fulfils promise of 600 community jobs'.

30 Ramaphosa was at the time executive chairman of Shanduka and also a member of the ANC executive committee.

31 'Platinum Wealth Holds no Shine for BaPo', *Mail and Guardian*, 29 June to 5 July 2012. See also Mail and Guardian online. http://mg.co.za/article/2012-11-23-00-baPo-community-wants-lonmin-right. http://mg.co.za/article/2012-12-14-00-lonmin-faces-community-showdown.

32 Interview with R.E. Maimane, by Bernard Mbenga and Andrew Manson, Wonderkop, 8 February, 17 February 2013.

33 *A Review of Platinum Mining in the Bojanala District of the North-West Province*, 2011. Bench Marks Foundation, p. 80.

34 There is another branch of the baKubung, the baRatheo, who live at Mathopestad. For information on their past, see H.L. Dugmore, 'Land and the Struggle for Sekama', pp. 43-48.

35 See *Mail & Guardian Online*, 18 December 2009; Mineweb, 27 August and 25 November 2010 *The Citizen*, 25 November 2010.

36 See Masilo, No v Bakubung ba Rantheo Traditional Council and Others, North West High Court, 23 September 2010.

37 Inteview with Lucas Moalusi, Bell Dewar Attorneys, Johannesburg, with Andrew Manson, 26 March 2011.

38 *Sunday World*, 16 December 2010. 'The Missing Shares - Skweyiya's Wife Named'.

39 *Sunday Times*, 21 March 2010. 'Royals Sling Mud at each other in War over R500 m.'

40 Bakgatla-ba-Kgafela Investment Profile, February 2013, p. 9; baKgatla ba Kgafela Traditional Administration Annual Review, May 2009-May 2010, p. 8.

41 'New director for Platmin', in *Mining Review.Com* http:www.miningreview.com/New/director/Platmin. Downloaded on 4 March 2013.

42 Ibid, p. 11.

43 Botsang Huma, 'Bakgatla partnership with the MQA improves Grade 12's Maths and Science results,' *Bua Kgabo*, Issue No. 1, January-September 2012, p. 4.

44 'Moruleng Mall set to bring great opportunities for its underserved communities," p. 1. http://www.sacommercialpropnews.co.za/south-africa-provincial-news/north-west-co downloaded on 4 March 2013.

45 See case no. 263/10, North-West High Court, Mafikeng, in the Matter between Nyalala Molefe John Pilane and the Traditional Council of the baKgatla ba Kgafela, and Pilane, M.K. (Moth) and Dintwe, R. Www.news24.com/SouthAfrica/News/Chief-in-R40mfraud.

46 *Sunday Times*, 18 March 2008.

47 Cited in A. Claassens, *Business Day*, 18 September 2012.

48 Cited in *City Press*, 8 December 2007.

49 Ibid.

50 See *Sunday World*, 12 December 2010, 'Chief Sold Us Out say BaKgatla Tribe'; *Sunday World*, 16 January, 2011, 'Royal Dispute Tribe Spurn Kgosi'.

51 Submission by Mrs Mary Mokgaetsi Pilane and Mr Mmothi Pilane of the BaKgatlaba Kautlwale to the Rural Development Portfolio Committee on the Repeal of the Black Authorities Act Bill. 21 July 2010.

52 LRC Update, 18 September 2012.//www.lrc.org.za/other-news/2315-2012-09-18-unaccountable-chiefs-are-a-recipe-for-a-new-marikana.

53 *Business Day*, 1 March 2013. 'Court Rejects Ban on Secession Talk'.

54 See North-West High Court, case no 263/10 in the Matter between Nyalala John Pilane (1st Applicant) and Traditional Council of the baKgatla ba Kgafela Traditional Community (2nd Applicant) and Mothi Pilane (1st Respondent) and Dintwe R, (2nd Respondent), Affidavit of Mmothi Pilane.

55 Submission by Mrs Mary Mokgaetsi Pilane and Mr Mmothibi Pilane of baKgatla ba Kautlwale, to Rural Development Portfolio Committee on the Repeal of the Black Authorities Act Bill.

56 Molopyane, 'Feature: Bakgatla will probe audit report,' in *The New Age Online*, p. 2.

57 BaKgatla ba Kgafela Traditional Administration Review 2012, p. 6.

58 See for example, the baKwena ba Mogopa in the Bethanie district whose affairs until recently were characterised by factionalism, legal wrangles, disputes over leadership and ineffective intervention by the NWP government. See Manson, 'Mining and Traditional Communities', pp. 417-419.

59 J. Carruthers, 'National Parks and Game Reserves, the Transvaal and Natal: pro-tected for the people or against the People.' Paper to 16th Biennial Conference of the South African Historical Society, 1997. p. 9

60 Ibid.
61 For a fuller account. See A. Manson and B. Mbenga, 'The Evolution and Destruction of *Oorlam* Communities in the Rustenburg District of South Africa: The cases of Welgeval and Bethlehem', *African Historical Review*, vol 4, (2009).
62 Keenan Report, p. 75.
63 Carruthers, 'National Game Parks...for the People or Against the people', p.10.
64 R. Davies, *A Description and History of Madikwe Game Reserve*, Madikwe Development Series, no1, North West Parks Board (Rustenburg, 1997), p.11.
65 Op. cit., p. 13.
66 D. Perkins, Madikwe Game Reserve: Baseline Survey and Preliminary Investigations into Attitudes towards the Reserve and Socio-economic Needs", Development Bank of South Africa, 1993.
67 R. Davies and M. Brett (eds), *Madikwe Game Reserve: A Decade of Progress*, (Rustenburg, 2003) Preface.
68 Ibid.
69 At least one community, the baTlokwa ba Kgosi, still have an outstanding claim on properties within the Park.
70 These findings are contained in E. Koch and P.J. Massyn, 'The Madikwe Initiative: A Programme Designed to Optimise Local Benefit by Integrating the Conservation of Wildlife with Local Economic Development', in Davies and Brett (eds), *Madikwe Game Reserve*. pp. 27-42.
71 Lyrics from song sung by Run-DMC, Kurtis Blow and Africa Bambaataa. Cited in T. Sannar, 'Playing Sun City: The Politics of Entertainment at a South African Mega-Resort', unpublished paper to Conference of South African Historical Society, Pretoria, 2009, p. 5.
72 Keenan Report, p. 36.
73 Interview with Dan Ntsala, General Manager, Cabanas Hotel, Sun City, with Bernard Mbenga, 16 March 2011.

Conclusion

From the imposition of white rule in the 1840s to the beginning of the new millennium, the African people in the North West Province experienced dramatic changes. Some of these would be readily apparent just from an observation of the regional environment. Apart from the obvious and conspicuous footprints of economic and social transformation (roads, railways, bridges, mission stations, churches, urban housing and telecommunications) one can note further features. First, the black population was forced to reside in reserves set aside for them, or on privately acquired 'tribal' farms, and many still live here or have a foot in these areas. The reserves, although still home to thousands of people, are for the most part barren and barely offer the chance of accumulation on any significant scale. Second, large tracts of land, mostly in the 'maize triangle' of the old western Transvaal, are still concentrated in the hands of commercial farmers. Third, the earth is scarred by the imprint of numerous mines, and the rivers and air polluted as a consequence. Finally, the region's wildlife has been herded into game parks, now a source of considerable profit. This book records and analyses a number of the key moments and developments in the human history of the North West Province.

During and following the so-called *difaqane*, a number of opportunistic, determined and well connected young men brought together the remnants of ethnically affiliated people and independent groups and began to reoccupy land in the bushveld and the thornveld area south of the Molopo River. To do so they had to negotiate with the unpredictable *trekboers* establishing themselves from the late 1830s around the available water sources of the region. This required astute leadership skills and the ability to negotiate on equal terms with the white authorities. These skills were exemplified by Moiloa of the baHurutshe. Historians have resorted to the frequent use of the terms 'collaboration' and 'resistance' to describe the reaction of African societies to colonial rule. While these terms still have relevance, the response to colonial rule of Africans such as Moiloa and all the significant nineteenth century baTswana leaders defy neat compartmentalisation into collaborators or resisters – there are many subtleties and ambiguities within and between these rather stark contrasts. Moiloa, in 1852, may have depicted himself as 'the dog of the boers' but he sought to free himself and his followers from this subjugation and was fairly successful in doing so, as were the other important African chiefs of the nineteenth century. In so doing they were able to retain a reasonable degree of political and national security.

The South African War (1899-1902) was a watershed moment for many of the black rural inhabitants of the bushveld, an opportunity to redress the wrongs of the past for those chiefdoms struggling to survive the worst excesses of life under the South African Republic. The baKgatla ba Kgafela living in Moruleng were unwaveringly antagonistic to the Boers, as is well illustrated by the battle at Derdepoort. There were also many similarities between the role and involvement of the baHurutshe in the Marico and the baKgatla in the Pilanesberg during the war: both became involved of their own volition but also as a consequence of the support and encouragement of the British military authorities; in both districts the involvement of the two communities was characterised by a general lawlessness which included the looting of Boer properties, in particular cattle. The baKgatla looted more than any other baTswana group, thus inviting an interpretation of the war as a 'peasant' revolt against a Boer agrarian ruling class. For the baRolong in Mafikeng the war provided the perfect opportunity to cement a closer relationship with the British, one that – for much of the last decades of the nineteenth century – proved to be rather vexed and contradictory.

Although the baKgatla and baHurutshe managed to loot Boer farms and rustle substantial livestock, the gains were largely temporary and the attempt to roll back years of dispossession was short-lived. The end of the war saw the British administration ensuring a quick restoration of law and order and a return to pre-war master-servant relations on the land between the African people of the bushveld and the local white producers.

Most of the Setswana speakers in the bushveld region and in the reserves were assured of land security in the decades after Union. The ability of many *merafe* outside the reserves to acquire more land, and the relative security of tenure in the reserves, placed a limit on the extent to which the inhabitants were incorporated into the country's political and economic structures.[1] In the Bechuanaland reserves and in Moiloa's Reserve, peasant producers and rural elites held their own and indeed there were periods when the residents experienced an improvement in wealth and living standards, but it was not an even process, as droughts and diseases could plunge households into impoverishment. As in many of South Africa's reserves, men and women looked outside the households for work but returned home to take up livelihoods based on subsistence agriculture and stock keeping. Their absorption into South Africa's mainstream capitalist economy was therefore uneven. At the same time, new sources of dissent arose among these rural communities in the first half of the twentieth century. Independently-minded households, often in collusion with disillusioned factions of the ruling families, sought to break the stranglehold that traditional authorities held over the material resources of the *merafe*. The authorities were forced to intervene, setting up a three-way interaction in which the state tried to balance the need for orderly governance with the application of customary law and practice. The result was that customary law, which was flexible and open to change, became increasingly fixed and codified, largely to the benefit of established chieftaincies. The leaders of these challenges to chiefly rule were not necessarily men and women who had entered into the wider urban labour market – many were rooted in the rural polities and economies within which they lived and worked. The factionalism and disruptions among the baFokeng and baKwena ba Mogopa the baKubung and baHurutshe were largely contained by the state, although the causes never simply evaporated and these squabbles foreshadowed very similar conflict among those people living in the platinum belt after the mining revolution – only then the stakes were to be markedly greater.

Among the baHurutshe in the 1950s, however, state intervention in intra-Hurutshe tensions between a rural elite and the traditional order, contributed to concerted opposition to the apartheid system, culminating in the revolt of 1957-58. This local resistance fed later into the broader national objectives of the ANC, and a firm foundation was laid for creation of the 'pipeline' into Botswana through which presidents Mandela, Mbeki and Zuma all exited the country.

The incorporation of most of the African population into the Tswana territorial authority and then Bophuthatswana in 1977 led to a troublesome two decades under the heel of the homeland's administration and its authoritarian president, Lucas Mangope. It was a time of dislocation and insecurity which also led to violence and loss of life. Several baTswana communities played an important and memorable role in the story of resistance to the homeland regime and the system of apartheid in general. Through stubborn resistance and a range of tactics, in particular recourse to the courts, they managed to retain a hold on the resources they valued and needed, and never fully surrendered their autonomy, land or other resources to the new authority.

The banning and exile of the liberation movements placed huge significance on the neighbouring frontline state of Botswana. The building of the crucial 'pipeline' between the two countries was made possible by the involvement and endurance of individuals living along the border. The rural baTswana were thus a key component in the building of a successful 'underground' in the region for the liberation movements – much of the historical trajectory of these communities was directed at liberation, from settler and from colonial regimes, from the apartheid system or from their own repressive leaders and representatives. There must have been cause for optimism when the Bophuthatswana regime dramatically fell in March 1994, and was replaced first with an interim administration and then – following the first democratic elections – with the provincial administration of the newly conceived North West Province, into whose boundaries all the bushveld communities were gathered.

During the Mangope years, politics and economic development became increasingly intertwined because of increased production and the vast profits realised from mining. The story of how the baFokeng took on Impala Platinum and the Bophuthatswana government is a memorable one. It was the first and very important step in the 'corporatisation' of the baFokeng; but it also had a wider significance for future relations between the mining companies and traditional communities in that it tilted the balance of power very much more towards the latter. All the *merafe* within the platinum belt were now potentially 'mineral-rich' and, under the new dispensation after 1994, were able to begin to negotiate with the mining companies as equal partners.

However, it emerges from the cases discussed in Chapter Seven that sharing the benefits of mining (from royalties, shares or employment) with local communities through their 'traditional authorities' is at best precarious and at worst disastrous. Every ethnic group holding mineral assets in the platinum mining fields of the province has experienced some form of economic or political turmoil.[2] The baFokeng, despite their undoubted achievements, are by no means exempt from the effects of this toxic mix. The critical issues can be summarised: the opportunities for misappropriation and corruption; questions surrounding the legitimacy and membership of such communities; objections posed, often by alleged independent factions within the communities who resented the concentration of power and control of mining assets in the hands of chiefs; and squabbles and divisions between different factions of the 'royal' families.[3] As the anthropologists John and Jean Comaroff observe, 'the more that ethnically defined populations move toward the model of the profit-seeking corporation, the more their terms of membership tend to become an object of concern, regulation and contestation'.[4] This would certainly apply to the baFokeng and, to a lesser extent, the baKgatla. Thus there has been a revival of regional and mineral-driven ethnic affiliation in the province and even beyond its borders in Botswana – belonging to a potentially rich community can be regarded as a probable passport to increased benefits and opportunities.

Compounding all these problems and insecurities, from 2011 there is the vexed issue of nationalisation of the mining sector and the rise of extreme social and economic tensions in the platinum belt that led to the Marikana shootings of August 2012. This tragic incident took place at Wonderkop on land belonging to the baPo ba Mogale. The spectre of nationalisation of the mining industry receded after the ANC, at its Mangaung elective conference in December 2012, ditched plans to forge ahead with the policy in favour of 'strategic state ownership'. Normality has by no means returned to the sector. In mid-January 2013, Amplats announced plans to close four shafts in its Rustenburg operations and lay off 14 000 workers. The unions and the government reacted with anger and indignation, and forced Amplats into a temporary suspension of its intentions. The Marikana commission of enquiry began to investigate the causes of the disaster in March 2013 and the baPo are set to give evidence to the commission to express the problems facing traditional communities caught in the cross-fire between the needs and objectives of workers and squatters and the mining industry and the state.[5] The future of Lonmin is uncertain and the company's policies and practices are under severe scrutiny. Competition and conflict between the two major unions, the National Union of Mineworkers (NUM) and the Association of Mineworkers and Construction Union (AMCU) from mid-2013 led to further disruption, and was responsible in many ways for fuelling the crippling strike action of January 2014. These generalised tensions sweep up all residents of the platinum belt in their vortex. The platinum mining sector was under threat by the beginning of 2013, with falling prices and production, and it remains to be seen how developments in the sector will affect the position of the African denizens of Rustenburg's platinum belt.

It is a truism applicable to most African societies, but land is the pivot on which the history of the Tswana in the North West Province rests. The vast majority of black people in the province resided on the land or had strong attachments to it, whether they were living in the reserves or on 'tribally' owned land. Even those living on white farms as labourers, tenants or sharecroppers retained some sense of their ethnic affiliation and land base; Kas Maine, for example, the protagonist in Charles van Onselen's biography of a sharecropper on farms in the southwestern Transvaal, sought sanctuary towards the end of his life with his Kubung family then living in Ledig.[6] Apart from Rustenburg itself, there was little urbanisation until the last quarter of the twentieth century, coinciding with the mining revolution, and migrants leaving rural districts headed mostly for the Witwatersrand. The land was tamed, fought over, possessed and repossessed, purchased and exchanged. It was the cause of internal conflict and intense resistance to the intrusive forces of colonialism and apartheid. Beyond its significance for material and political security, land was a source of comfort, patrimony and identity.

Finally, for many of its inhabitants, it proved to be the fount of unanticipated mineral wealth.

ENDNOTES

1   W. Beinart, 'Beyond Homelands: Some Ideas about the History of Rural Areas in South Africa', *South African Historical Journal*, vol. 64, no 1, (2012). p. 6
2   In particular the baKubung ba Ratheo, who live around Ledig, and the baKgabatla ba Kgafela, in the foothills of the Pilanesberg range.
3   Manson, 'Mining and Traditional Communities', p. 423.
4   Comaroff and Comaroff, *Ethnicity Inc*, p. 65.
5   Personal communication, O. Kgomo Mafikeng, 21 January 2013. Kgomo, Mokhetle and Tlou are the attorneys for the baPo.
6   C. van Onselen, *'The Seed is Mine': The Life of Kas Maine, A South African Sharecropper*, 1894-1985 (Cape Town: David Philip, 1996), p. 472.

Bibliography and Sources

UNPUBLISHED PRIMARY SOURCES

Official

Transvaal Archives (TA)

SS Archives of the State-Secretary 1851-1899. Incoming, outgoing and supplementary correspondence and reports.
SN Archives of the Superintendant van Naturelle Sake, 1877-1899. Correspondence, Reports and Commissions.
SNA Archives of the Secretary for Native Affairs, 1910-1948. Correspondence and Reports.
KG Archives of the Kommandant-Generaal. 1881-1899.
LD Archives of the Law Department, 1910-1938. Correspondence and Reports.
BAO Archives of the Department of Bantu Administration and Development, Select for 1950s.
JUS Archives of the Secretary of Justice.
LC Landros Correspondence for Marico, Pilanesburg and Rustenburg, 1875-1899.
NTS Archives of the Native Affairs Department, 1920-1958.

Botswana National Archives (BNA)

Correspondence of the Commissioner for Bechuanaland Protectorate.
RC Correspondence of the Resident Commissioner for Bechuanaland.

Cape Archives Depot (CAD)

Archives of the Secretary for Native Affairs Vryburg and Mafikeng.
Magisteral Records for Vryburg and Mafikeng.

Other Archival Depositaries

J.H.E. Hamm Collection, North-West University, Potchefstroom, F. Postma Library, Western Transvaal History Archive.
University of the Witwatersrand, William Cullen Library, Historical and Literary Papers Collection.
University of South Africa Archival Collections.

MISSIONARY RECORDS AND PAPERS

London Missionary Society (LMS). Incoming Letters and Reports, primarily for Marico District, 1846-1852.

Archives of the Evangelishe-Lutherisches Missionwerk in Niedersachsen (ELM), incorporating records of the Hermannsburg Missionary Society, (HMS). Records available in Hermannsburg, and at the University of South Africa, Hesse Collection.

Hermannsberg Missions Berichte (HMB) Printed Annual Reports of the HMS.

UNPUBLISHED REPORTS

Keenan Report into the Establishment of the Pilanesberg Game Reserve. Commissioned by the Bophuthatswana Parks Board, 1984.

PRINTED PRIMARY SOURCES

Transvaal Colony
Annual Reports of the Transvaal Native Affairs Department, 1902-1910.

Report of the Commissioner for Native Affairs Relative to the Acquisition and Tenure of Land by Natives in the Transvaal (Pretoria, 1904)

Native Affairs Deprtment (TNAD), *Short History of the Native Tribes of the Transvaal* (Pretoria, 1905).

TNAD *The Laws and Regulations Specially Relating to the Native Population of the Transvaal* (Pretoria, 1907).

War Office General Staff, *The Native Tribes of the Transvaal* (London, 1905).

Union of South Africa
UG 19-1916. Report of Natives' Land Commission.

UG 23-1918. Report of Natives' Land Commission, Western Transvaal.

Other published reports
BaKgatla ba Kgafela Investment Profile, February 2013, *BaKgatla ba Kgafela Traditional Administration Annual Review*, May 2009.

Evidence taken at Bloemhof before the Commission appointed to investigate claims … to the Diamond Fields (The Bloemhof Blue Book). CA 21/1.

NorthWest Province Gazette, Schedule One, Office of the Premier, 21 September 2007.

Perkins, D., Madikwe Game Reserve: *Baseline Survey and Preliminary Investigations into Attitudes towards the Reserve and Socioeconomic Needs*, Development Bank of Southern Africa, 1993.

Review of Platinum Mining in the Bojanala District of the North West Province, 2011. Bench Marks Foundation.

Royal BaFokeng Nation, *Population and Use of Land Audit*, (PULA) Study, 2010-2011.

Court Cases

North-West High Court, case no 263/10 in the Matter between Nyalala John Pilane (1st Applicant) and Traditional Council of the baKgatla ba Kgafela Traditional Community (2ndApplicant) and Mothi Pilane (1st Respondent) and Dintwe R, (2nd Respondent), Affidavit of Mmothi Pilane.

Mafikeng High Court, North-West Province, no. 999/2008, In the Matter between the Royal baFokeng Nation, Applicant and the Minister for Rural Development and Land Reform (1st Respondent).

Submission by Mrs Mary Mokgaetsi Pilane and Mr Mmothibi Pilane of baKgatla ba Kautlwane, to Rural Development Portfolio Committee on the Repeal of the Black Authorities Act Bill. North-West High Court, case no 263/10 in the Matter between Nyalala John Pilane (1st Applicant) and Traditional Council of the baKgatla ba Kgafela Traditional Community (2nd Applicant) and Mothi Pilane (1st Respondent) and Dintwe R, (2nd Respondent), Affidavit of Mmothi Pilane.

North Gauteng High Court, Pretoria, Regional Authority of the BaPo Ba Mogale v Kenoshi and the Premier, North West Province, Case no 31876-10 2010, July 2011.

CONTEMPORARY PUBLICATIONS

Amery, L.S., *The Times History of the War in South Africa*, vol. III (London, 1905).

Campbell, J., *Travels in South Africa Undertaken at the Request of the Missionary Society Narrative of a Second Journey*, 2 Vols. (London: Westley, 1822).

Chapman, J., *Travels in the Interior of South Africa, Comprising Fifteen Years Hunting and Trading*, (London, 1868).

Davitt, M., *The Boer Fight for Freedom* (London: Funk and Wagnalls, 1902).

Delegorgue, A., *Travels in South Africa*, 2 vols. (Scottsville: Natal University Press, 1997). Translated by F. de B. Webb.

Freeman, J.J., *A Tour to South Africa* (London, 1851).

Holub, E., *Seven Years in South Africa, Researches and Hunting Adventures between the Diamond-Fields and Zambesi, 1872-1879*, 2 vols. (London: Sampson Low, 1881).

Kay, S., *Travels and Researches in Caffraria* (London, 1833).

Knight, E.F., *South Africa after the War: A Narrative of Recent Travel* (London: Longman, 1903).

Livingstone, D., *Missionary Travels and Researches in South Africa* (London: John Murray, 1857).

Lovett, R., *The History of the London Missionary Society, 1795-1895* (London: 1899).

Mgadla, P. and Volz, C. (eds), *Words of Batswana: Letters to Mahoko a Bechwana, 1883-1896* (Cape Town: Van Riebeeck Society, 2006).

Moffat, R., *Missionary Labours and Scenes in South Africa* (London, 1885).

Wilson, D.M., *Behind the Scenes in the Transvaal* (London: Cassel, 1901).

PUBLISHED SECONDARY SOURCES

Books

Agar-Hamilton, J.A.I., *The Native Policy of the Voortrekkers: An Essay in the History of the Interior of South Africa, 1836-1858* (Cape Town: Maskew Miller, 1928).

Agar-Hamilton, J.A.I., *The Road to the North, South Africa, 1852-1886* (London: Longman, 1937).

Anderson D.M and Killingray, D., (eds), *Policing the Empire: Government, Authority and Control, 1830-1980* (Manchester: Manchester University Press, 1991).

Barmmann, H.W., *Tiragalo E Khutshwana Ka Bafurutshe, Ba Ba Dinokana LeBakwena Ba BaFokeng Le Bakwena Ba Ga Mogopa Ba Bethanie,* (Rustenburg: Evangelical Lutheran Church in Southern Africa, 2005).

Barmmann, H.W., *130th Anniversary, 1866-1996, Evangelical Lutheran Church of Southern Africa, Saron Parish, Phokeng* (Rustenburg: Evangelical Lutheran Church, 1996).

Beinart, W. and Bundy, C., *Hidden Struggles in Rural South Africa: Politics and Popular Movements in the Transkei and the Eastern Cape, 1890-1930* (Johannesburg: Ravan, 1987).

Beinart,W., *The Political Economy of Pondoland, 1860-1930* (Cambridge: Cambridge University Press, 1982).

Bergh, J. and Morton F., (eds), *'To Make them Serve ...' The 1871 Transvaal Commission on African Farm Labour* (Pretoria: Protea, 2003).

Black Sash, *Grasping the Prickly Pear: The Bophuthatswana Story* (Johannesburg, Black Sash, 1990).

Bizos, G., *Odyssey to Freedom* (Johannesburg: Random House, 2007).

Bozzoli, B. with Nkotsoe, M., *Women of Phokeng: Consciousness, Life Strategy and Migrancy in South Africa, 1900–1983* (Johannesburg: Ravan, 1991).

Bradford, H., *A Taste of Freedom: The ICU in Rural South Africa, 1924-1930* (New Haven and London: Yale University Press, 1987).

Breutz, P-L., *The Tribes of the Rustenburg and Pilanesberg Districts* (Native Affairs Department, Ethnological Publication no. 28, Pretoria, 1953).

Breutz, P-L., *The Tribes of the Marico District* (Pretoria: Native Affairs Department, Government Printer, Ethnological Publication no. 31, 1957).

Breutz, P-L., *Tribes of the Mafeking District,* (Pretoria: Native Affairs Department, Government Printer, 1957).

Breutz, P-L., *Tribes of the Vryburg District* (Pretoria: Native Affairs Department, Government Printer, 1959).

Breutz, P-L., *History of the Batswana and the Origins of Bophuthatswana* (Dr. P.L. Breutz, Ramsgate, 1987).

Carruthers, V., *The Magaliesberg* (Johannesburg: Southern, 1990).

Coertze, R.D., *BaFokeng Family Law and Law of Succession* (Pretoria: SABRA,1990).

Comaroff, J., *Of Revelation and Revolution: Christianity, Colonialism and Consciousness in South Africa* (Chicago: Chicago University Press, 1997).

Comaroff, J.L. and Comaroff, J., *Ethnicity Inc.* (Scottsville: University of Kwazulu Natal Press, 2009).

Cuthbertson G., Grundlingh, A. and Suttie M.L.,(eds), *Writing a Wider War: Rethinking Gender, Race and Identity in the South African War, 1988-1902* (Cape Town and Athens: David Philip and Ohio University Press, 2002).

Davenport, T.R.H. and Hunt, K.S. (eds), *The Right to the Land* (Cape Town: David Philip, 1974).

Davies, R., *A Description and History of the Madikwe Game Reserve* (Rustenburg: North West Parks Board, 1997).

Davies, R. and Brett, M., *Madikwe Game Reserve: A Decade of Progress* (Rustenburg: North West Parks Board, 2003).

Delius, P., *The Land Belongs to Us: The Pedi Polity, the Boers and the British in the Nineteenth-Century Transvaal* (Johannesburg: Ravan, 1983).

Delius, P., *A Lion Among the Cattle: Reconstruction and Resistance in the Northern Transvaal* (Johannesburg: Ravan, 1996).

Du Pisani, K., *The Last Frontier War, Braklaagte and the Struggle for Land* (Pretoria: Rosenburg Publishers and Unisa Press, 2009).

Eldredge, E.A. and Morton, F., (eds), *Slavery in South Africa: Captive Labor on the Dutch Frontier* (Boulder and Scottsville: Westview and University of Natal Press, 1994).

Etherington, N., *The Great Treks: The Transformation of Southern Africa, 1815-1851* (London: Longman, 2005).

Hamilton, C., (ed.), *The Mfecane Aftermath: Reconstructive Debates in Southern African History* (Johannesburg and Scottsville: Witwatersrand University Press and University of Natal Press, 1995).

Hamilton, C., Mbenga, B. and Ross, R., *The Cambridge History of South Africa, From Early Times to 1885*, vols 1 and 2 (Cambridge: Cambridge University Press, 2010 and 2012).

Hasselhorn, F., *Mission, Land Owernship and Settlers' Ideology* (Johannesburg: South African Council of Churches, 1987).

Hooper, C. *Brief Authority* (London: Penguin, 1960).

Jacobs, N.J., *Environment, Power and Justice: A South African History* (Cambridge: Cambridge University Press, 2003).

Jeffrey, A., *Conflict at the Crossroads in Bophuthatswana* (Johannesburg: South African Institute of Race Relations, 1992).

Keegan, T., *Rural Transformations in Industrialising South Africa, The Southern Highveld to 1914* (Johannesburg: Ravan, 1986).

Kistner, W., *The Anti-Slavery Agitation against the Transvaal Republic, 1852-1862* (Parow: Archives Year Book, 1952).

Klein, M., (ed.), *Peasants in Africa, Historical and Contemporary Perspectives*, (Beverly Hills: Sage, 1980).

Krikler, J., *Revolution from Above, Rebellion from Below, The Agrarian Transvaal at the Turn of the Century* (Oxford: Clarendon, 1993).

Landau, P., *Popular Politics in South Africa, 1400-1948* (Cambridge: Cambridge University Press, 2010).

Legassick, M.C., *The Politics of a South African Frontier: The Griqua, the Sotho-Tswana and the Missionaries, 1780-1840* (Basel, 2010).

Limb, P., *The ANC's Early Years: Nation, Class and Place in South Africa before 1940* (Pretoria: Unisa Press, 2010).

Limb, P., Etherington N. and Midgley, P. (eds), *'Grappling with the Beast': Indigenous Southern African Response to Colonialism, 1840-1930* (Leiden and Boston: Brill, 2010).

Lodge, T., *Black Politics in South Africa from 1945*, (Johannesburg: Ravan, 1983).

Lonsdale, J. (ed.), *South Africa in Question* (Cambridge: Cambridge University Press, James Currey and Heinemann, 1988) .

Makgala, J., *History of the BaKgatla ba ga Kgafela in Botswana and South Africa* (Pretoria: Crink City Publications, 2009).

Mbenga, B. and Manson, A., *People of the Dew: A History of the BaFokeng of the Rustenburg District, South Africa, from Early Times to 2000* (Johannesburg: Jacana, 2010).

Mokgatle, N., *The Autobiography of an Unknown South African* (Berkeley: University of California Press, 1971).

Molema, S.M., *Montshiwa, BaRolong Chief and Patriot* (Cape Town: Struik, 1966).

Morton, F., *When Rustling Became an Art: Pilane's Kgatla and the Transvaal Frontier, 1820-1902* (Cape Town: David Philip, 2009).

Pakenham, T., *The Boer War* (Johannesburg: Jonathan Ball, 1979).

Rasmussen, R.K., *Migrant Kingdom: Mzilikazi's Ndebele in South Africa* (London: Rex Collings, 1978).

Rumley, D. and J. Minghi (eds), *The Geography of Border Landscapes* (London: Routledge, 1991).

Schapera, I., *A Handbook of Tswana Law and Custom* (London: Oxford University Press, 1938).

Schapera, I., *The Ethnic Composition of Tswana Tribes* (London School of Economics Monograph, 1952).

Schapera, I., *A Short History of the BaKgatla-ba ga-Kgafela of the Bechuanaland Protectorate* (Cape Town, 1942).

Schapera, I., (ed.), *Ditirafalotsa Merafe ya Batswana* (Alice: Lovedale Press, 1940).

Schapera, I., with Comaroff, J.L., *The Tswana* (revised edition) (London: Kegan Paul, 1975).

Schutte, G.J., *The Sons of Magato (Mokgatle Thethe, Kgosi of the BaFokeng) in Delft, Holland* (Amsterdam: South African Institute, 2007).

Shillington, K., *The Colonisation of the Southern Tswana, 1870-1900* (Johannesburg: Ravan, 1985).

Shillington, K., *Luka Jantjie: Resistance Hero of the South African Frontier* (London and Johannesburg: Aldridge Press and Wits Unversity Press, 2011).

South African Democracy Education Trust (SADET), *The Road to Democracy in South Africa*, Vols 1 and 2 (Pretoria: Unisa Press, 2006, 2009).

Swanepoel, N., Esterhuysen, A. and Bonner, P. (eds), *Five Hundred Years, Rediscovered* (Johannesburg: Wits University Press, 2008).

Tema, B., *The People of Welgeval*, (Cape Town: Zebra, 2005).

Thompson, L., (ed.), *African Societies in Southern Africa* (London: Heinemann, 1969).

Van Onselen, C., *"The Seed is Mine": The Life of Kas Maine, a South African Sharecropper, 1894-1985* (Cape Town: David Philip, 1996).

Van Warmelo, N.J., *A Preliminary Survey of the Native Tribes of South Africa* (Pretoria: Native Affairs Department, Ethnological Publication no. 5, 1935).

Warwick, P., *Black People and the South African War, 1899-1902* (Johannesburg: Ravan, 1983).

Wells, J., *We Now Demand, The History of Women's Resistance to Pass Laws in South Africa* (Johannesburg: Wits University Press,1994).

Wulfsohn, L., *Rustenburg at War: The Story of Rustenburg and its Citizens in the First and Second Anglo-Boer Wars* (Rustenburg: L.M. Wulfsohn, 1987).

Journals

Ake, C., 'What is the Problem of Ethnicity in Africa?' *Transformation*, 22. (1993).

Beinart, W., 'Beyond Homelands: Some Ideas about the History of Rural Areas in South Africa', *South African Historical Journal*, 64, 1 (2012).

Bergh, J., "We must never forget where we come from": The BaFokeng and their land in the nineteenth century Transvaal, *History in Africa*, 32 (2005).

Botha, H. J., 'Die Moord op Derdepoort, 25 November 1899. Nie Blankes in Oorlogsdiens', *Militaria*, 1 (1969).

Comaroff, J.L., 'Chiefship in a South African homeland: A case study of the Tshidi chiefdom of Bophuthatswana', *Journal of Southern African Studies*, (October 1974).

Cook, S.E., 'Chiefs, kgosis, corporatization, and democracy: A South African case study', *Brown Journal of World Affairs*, 12, 2 (2005).

Cook, S.E., 'The Business of Being BaFokeng: The Corporatisation of Tribal Authority in South Africa', *Current Anthropology* 52 (suppl 3) S151-S159 .

Delius, P. and Trapido, S., '*Inboekselings* and *Oorlams*: The Creation and Transformation of a Servile Class', *Journal of Southern African Studies*, 8, 2 (1982).

Drummond, J. and Manson, A., 'The Rise and Demise of Agricultural Production in Dinokana Village, Bophuthatswana', *Canadian Journal of African Studies*, 27, 3 (1993).

Dubow, S., 'Holding a just balance between white and black: The Native Affairs Department in South Africa c.1920–1933', *Journal of Southern African Studies*, 12, 2 (April 1986).

Dugmore, H., 'The Rise to Power of the Mmonakgotla Family of the Bakubung', *Africa Perspective*, nos 2 and 3 (1987).

Ellenberger, J., 'The Bechuanaland Protectorate and the Boer War,' *Rhodesiana*, XI (1964).

Grobler, J., 'Van Viljoen, The South African Republic and the Bakwena, 1848-1865,' *South African Historical Journal*, 36, (1997).

Jacobs, N. J., 'The Great Bophuthtswana Donkey Massacre: Discourse on the Ass and the Politics of Class and Grass', *American Historical Review*, 108 (1997).

Jensen, F.H.W., Note on the Bahurutshe', *African Studies*, 4 (1957).

Lawrence, M. and Manson, A., 'The "Dog of the Boers": The Rise and Fall of Mangope in Bophuthatswana'. *Journal of Southern African Studies*,20 (1994).

Lestrade, G.P., 'Some notes on the political organisation of the Bechwana', *South African Journal of Science*, 25 (December 1928).

Lissoni, A., "Chieftaincy and resistance politics in Lehurutshe in the apartheid era", *New Contree*, 67 (2013).

Manson, A., 'The Hurutshe Resistance in the Zeerust District of the Western Transvaal, 1954-1957', *Africa Perspective*, 22 (1983).

Manson, A., 'The Hurutshe and the formation of the Transvaal state, 1835–1875', *International Journal of African Historical Studies*, 25, 1 (1992).

Manson, A.H., 'Mining and "Traditional Communities" in South Africa's "Platinum Belt": Contestations Over Land, Leadership and Assets in North-West province, c. 1996-2012, *Journal of Southern African Studies* 39, 2, (2013).

Manson, A. and Mbenga, B., '"The Richest Tribe in Africa": Platinum Mining and the BaFokeng in South Africa's North-West Province,' *Journal of Southern African Studies*, 29, 1 (2003).

Manson, A. and Mbenga, B., 'The Evolution and Destruction of *Oorlam* Communities in the Rustenburg District of South Africa: The Cases of Bethlehem and Welgeval, 1850s-1980', *African Historical Review*, 41, 2 (2009).

Manson, A, and Mbenga, B., 'Bophuthatswana and the North-West Province: from Pan-Tswanaism to Mineral based Ethnic Assertiveness', *South African Historical Journal*, 64,1 (2012).

Manson, A, and Mbenga, B., 'The African National Congress in the Western Transvaal/ northern Cape Platteland, c. 1910-1965', *South African Historical Journal*, 64, 3 (2012).

Matthews, Z.K., 'A Short History of the Tshidi-BaRolong', *Fort Hare Papers*, (June 1945).

Mbenga, B., 'Forced Labour in the Pilanesberg: The Flogging of Chief Kgamanyane by Commandant Paul Kruger, Saulspoort, April 1, 1870', *Journal of Southern African Studies*, 27, 1 (1997).

Mbenga, B and Morton, F., 'The Missionary as Broker: The Rev. Henri Gonin, the BaKgatla and the South African Republic,'(1997) *South African Historical Journal*, 36.

Morton, R.F., 'Slave Raiding and Slavery in the Western Transvaal after the Sand River Convention', *African Economic History*, 20 (1992).

Morton, R.F., 'Linchwe 1 and the Kgatla Campaign in the South African War, 1899-1902', *Journal of African History*, 26 (1985).

Nasson, B., 'Doing Down their own Masters: Africans, Boers and Treason in the Cape Colony, 1899-1902'. *Journal of Imperial and Commonwealth Studies*, 12, 1 (1983).

Phimister, I., 'Rethinking the Reserves: Southern Rhodesia's Land Husbandry Act Reviewed', *Journal of Southern African Studies*, 19, 2, (1993).

Proske, W., 'The political significance of the early Hermannsburg mission in Botswana: An assessment of its role among Batswana, the British and the Boers', *Botswana Notes and Records*, 22 (1990).

Trapido, S., 'Aspects in the Transition from Slavery to Serfdom: The South African Republic, 1842–1902', *Collected Seminar Papers on the Societies of Southern Africa in the 19th and 20th Centuries*, vol. 6, no. 20 (May 1975).

UNPUBLISHED THESES AND PAPERS

Caldwell, S., Genealogies of change: Tradition and history in negotiations of political change in the Royal BaFokeng Nation, Senior Essay, Yale University, 2002.

Capps, G.J., Tribal Landed Property: The Political Economy of the BaFokeng Chieftaincy, South Africa, 1837-1994, D. Phil thesis, London School of Economics and Political Science, 2010.

Carruthers, J., National Parks and Game Reserves, the Transvaal and Natal: Protected for the People or Against the People, Paper to Conference of the South Africa Historical Society.

Cobbing, J., The Ndebele under the Khumalo 1820-1896, Ph.D thesis, University of Lancaster, 1976.

Comaroff, J., Competition for Office and Political Processes amontg the Barolong boo Ratshidi of the South African-Botswana Borderland, Ph.D thesis, University of London, 1973.

Drummond, J.H., Changing Patterns of Land Use and Agricultural Production in Dinokana Village, Bophuthatswana, (M.A. thesis, University of the Witwatersrand, 1992).

Dugmore, H.L., Land and the Struggle for Sekama: The Transformation of a Rural Community, The Bakubung of the Western Transvaal, (B.A Hons dissertation, University of the Witwatersrand, 1985).

Hasselhorn F., Mission and Land in Southern Africa, Illustrated by the Example of the Hermannsburg Mission in the Transvaal, unpublished paper, University of Gottingen, 1985.

Kriel, I., Visible through ethnicity: The baFokeng and their wealth, unpublished paper, Dept of Anthropology, University of Pretoria, 2008.

Manson, A., The Hurutshe in the Marico District of the Transvaal, 1848–1914, Ph.D thesis, University of Cape Town, 1990.

Mbenga, B.K., The baKgatla ba Kgafela in the Pilanesberg District of the Western Transvaal from 1899 to 1931, D.Litt et Phil. thesis, University of South Africa, 1997.

Mnwane, S. Participation and Paradoxes: Community Control of Mineral Wealth in South Africa's Royal baFokeng and baKgatla ba Kgafela Communities. (D.Phil. dissertation, University of Fort Hare, 2011).

Mohlamme, J. S., The role of black people in the Boer Republics during and in the aftermath of the South African War of 1899 -1902 , Ph.D. thesis, University of Wisconsin-Madison (1985).

Morton, B.C., A Social and Economic History of a Southern African Reserve: Ngamiland, 1898-1966, D. Phil. Thesis, Indiana University, 1966.

Ngcongco, L., Aspects of the History of the Bangwaketse, up to 1910, Ph.D thesis, Dalhousie University, 1976.

Parsons, Q.N., Khama III, the Bamangwato and the British, Ph.D thesis, University of Edinburgh, 1977.

Ramaroka, M.D., The Early Internal Politics of the BaRolong in the District of Mafikeng: Intra-Batswana Ethnicity and Political Culture from 1852-1920, M.A. thesis, North-West University, 2003.

Ramsay, F.J., The rise and fall of the Bakwena dynasty of south-central Botswana, 1820-1940, Ph.D thesis, Boston University, 1991.

Reddy, R.M., "Darlings No More": Tswana Mineworkers and Labour Unrest on the Platinum Mines of Bophuthatswana, B.A. (Hons) essay, University of the Witwatersrand, 2011.

Relly, G.U., The transformation of rural relationships in the western Transvaal, 1900–1930, M.A. dissertation, School of Oriental and African Studies, University of London, 1978.

Sanner T., "Playing Sun City": The Politics of Entertainment in a South African Mega-Resort, Paper to Conference of the South African Historical Society, Pretoria, 2009.

September, L., The Social Dimensions of Mining: Expectations and Realities in a Mining Induced Relocation, M.A. dissertation, University of the Witwatersrand, 2010.

Simpson, G.N., Peasants and politics in the western Transvaal, 1920-1940, M.A. dissertation, University of the Witwatersrand, 1986.

ORAL SOURCES

Kenneth Mosinyi, Dinokana, 18 January 1983, South African Institute of Race Relations Oral Archive.

Joanna Pule, Dinokana, 3 December 1982, South African Institute of Race Relations Oral Archive.

Mme Semane Molotlegi, Phokeng, 11 March 1999.

Cecil Tumagole, Phokeng, 18 January 1999.

James Sutherland, Johannesburg 3 August 1999.

Lucas Moalusi, Johannesburg, 26 March, 2011.

Dan Ntsala, 18 March 2011.

R. E. Maimane, Wonderkop, 8 February, 17 February 2013.

R. Ramolope and S. Molope, joint interview, with Bernard Mbenga, Saulspoort, 17 February 1994.

T.S. Pilane, interview, Saulspoort, 29 May 1993; B.N.O.Pilane, interview, Saulspoort, 8 October 1993.

Index

ba Mmanaana 20, 23–24
ba Sefikile 157–158
mining-tourism driven revenues
126–127, 144, 155–162
role in South African War 42–59,
82, 171
baKubung 9, 13
baMathope 34–35, 79, 81–82,
87n62, 131
baMonnakgotla 9, 34–35, 79–80, 82,
131, 150
baKubung ba Ratheo 9, 131, 167n34
Concerned Group 152–154
feud over assets 151–154
baKwena 9, 23, 27, 37, 50
baMogopa 12, 27, 72, 74–79, 98,
131–132, 168n58, 172
Setshele's 24–26, 33
baNgwaketse 4, 29, 32–34, 101
bantu authorities 114, 120
Bantu Authorities Act 89, 108n54,
119
see also Native Affairs Department
bantustan system 2, 13–14, 89, 91, 105,
114, 117–118, 124–125, 127–
128, 130, 135–136, 161, 165 *see
also* Bophuthatswana
baPhalane 27, 50, 134
baPo ba Mogale 7, 146–147, 149–150, 174
'D' account 146, 148, 167n15
Unemployment Forum 149
baRolong 2, 9, 11, 13, 25, 34, 92, 101–102
Barolong National Council (BNC)
96
boo raTlou boo Mariba 93, 95, 99
Rapulana 22, 93, 95–96, 99
Ratshidi 3, 11, 21, 57, 93, 95–99,
101–103, 105
role in South African War 42,
57–58, 171
baTlhaping 2–4, 9, 11, 13, 16, 33–34,
92–93, 134
baTlharo 2–3, 93–94, 96–97, 99
Bechuanaland Protectorate 3, 6, 44, 46,
53, 56, 116, 124, 171
Mochudi 29, 37–38, 43–50, 82–84,
158, 160
see also Botswana
Bechuanaland reserves 2, 12, 31, 65, 89,

91–95, 99–102, 103–105, 116,
170–171
Ganyesa 93–94, 101
as labour reservoirs 90, 102, 105, 273
Molopo 16, 93, 99, 101, 105
Morokweng 93, 99
Setlagole 92, 95–96, 99, 101–102,
105
white/black frictions in 92, 94
Behrens, W 27, 31, 75
Bell, Dewar and Hall 130, 136
Bethanie 27, 74–76
Bizos, George 111–112, 114
black economic empowerment (BEE) 14,
143, 149
Black Sash 130, 133
Bloemhof Commission 33–34
Boer contestation with African population
7, 12, 22
attack on Dimawe 23–26
during South African War 43–52,
54–58
Boer Republic *see* South African Republic
Boer society 1, 9, 12, 30–31
demand for labour/grazing 21,
23–24, 29, 35, 37
see also indentured/forced labour
Boer trekkers 7, 11, 19–22, 25, 29, 33, 170
bonded labour system *see* indentured
labour
Bophuthatswana 1, 3, 13, 91, 114, 121,
124–126, 135–139, 161–162–165,
172–173
attempted coup 136
Defence Force massacre of donkeys
134–135
fall of 130, 138, 173
forced incorporation into 127–133,
172
illusion of independence 125, 127,
165
National Seoposengwe Party
125–127
see also Pilanesberg game reserve;
Sun City resort
Bosman, Herman Charles 9, 142
Botha, Pik 131, 133
Botswana 3–4, 6, 23–24, 27, 29, 37, 54,
90, 92, 103, 110, 112–113,